Fundamentals of Daily Shop Floor Management

Survival and thriving in today's business environment require companies to continuously strive for operational excellence at all levels of the organization. Simply working to maintain existing operations is not an adequate or sustainable business strategy, especially when competing in a global market. To remain relevant, companies must adopt a process control and continuous improvement mentality as an integral part of their daily work activities. These two operational disciplines form the foundation and stepping stones for manufacturing excellence. Processes must be stable, capable, and controlled as a prerequisite for sustainable improvement. Sustainable improvements must be strategic, continuous, and focused on process optimization.

Modern-day manufacturing is rapidly changing in the face of technological, geopolitical, social, and environmental developments. These challenges are altering the way we think and act to transform raw materials into finished goods. Meeting these challenges requires particular attention to how we develop and engage people and apply technology for long-term sustainability and competitive advantage. This book takes you on a journey to explore the fundamental elements, management practices, improvement methods, and future direction of shop floor management.

Part 1 of this five-part manuscript considers workplace culture, organizational structure, operational discipline, and employee accountability as the foundation for a robust manufacturing system. Part 2 studies the impact of process standardization, data analytics, information sharing, communication, and people on daily shop floor management. Once the management system has been adequately described, Part 3 concentrates on its effective execution, monitoring, and control with a deep look into the people, methods, machines, materials, and environment that make it possible. Like every good manufacturing text, efficiency and productivity are key topics. That's why Part 4 explores various methods, tools, and techniques associated with product and process development, productivity improvement, agile methods, shop floor optimization, and manufacturing excellence. The final section, Part 5, shifts focus to emerging technologies, engaging the reader to contemplate technology's impact on the digital transformation of the manufacturing industry.

Fundamentals of Daily Shop Floor Management
A Guide for Manufacturing Optimization and Excellence

Philip J. Gisi

Routledge
Taylor & Francis Group

A PRODUCTIVITY PRESS BOOK

First Published 2023
by Routledge
605 Third Avenue, New York, NY 10158

and by Routledge
4 Park Square, Milton Park, Abingdon, Oxon, OX14 4RN

Routledge is an imprint of the Taylor & Francis Group, an informa business

ISBN: 978-1-032-37055-2 (hbk)
ISBN: 978-1-032-37054-5 (pbk)
ISBN: 978-1-003-33509-2 (ebk)

DOI: 10.4324/b23307

Typeset in Garamond
by SPi Technologies India Pvt Ltd (Straive)

Contents

PART IV   SHOP FLOOR IMPROVEMENTS

Preface

Early in 2021, I had a conversation with the Quality Manager of a small manufacturing facility in Wisconsin when I realized the need for this book. The plant was about to triple their sales due to the closing of a "sister" plant, with all of the closing facility's production lines being transferred to the Wisconsin facility over the next 12 months. This change was a rebirth for a facility on the verge of closing 6 months earlier due to a major loss of customer business. Before this loss, the facility worked at a comfortable pace under a friendly and relaxed work environment. This was about to change. They were taking on over 25 additional product lines, of varying types, sizes, and complexity. Plant floor space was going to be at capacity after the move and the support staff was not growing significantly with the increasing workload. These conditions were necessary to remain competitive. In short, their world was about to change and change quickly.

I am telling this story because during my conversation with this manager, I realized that he did not have a strong grasp of what it takes to run an efficient manufacturing operation in a competitive working environment. In the past, urgency to address problems quickly, efficiently, and permanently was buffered by sufficient resources, flexible production schedules, and adequate production capacity. The new norm of increased product volume, mix, and complexity demanded a change in employee mindset for long-term survival. They needed to adopt a more lean and agile approach to managing their daily shop floor activities and continuously improve their productivity to maintain a competitive market position. Based on my previous experience working with this team, they needed a deeper understanding of their approach to shop floor management relative to what's required to run a modern-day manufacturing operation in a high-cost country, competing in a global marketplace. My response was to coach the team as best I could during the transition while, in parallel, writing this book for those who may experience a similar situation in the future.

My first book *Sustaining a Culture of Process Control and Continuous Improvement: The Roadmap for Efficiency and Operational Excellence* [1] took a more strategic approach to lean manufacturing. It was a deep dive into the process maturity lifecycle and the steps to achieve, maintain, and improve lean performance on the road to operational excellence. See Figure P.1 which outlines the steps to operational excellence.

This book explores the more focused world of **daily shop floor management** and the on-going activities necessary to maintain process control and drive sustainable continuous improvements. It outlines a roadmap for manufacturing optimization and

Figure P.1 Steps to operational excellence [1].

excellence from a shop floor perspective. The roadmap is built on a controlled and continuously improving process that strives to increase operational efficiency and optimize product value to stakeholders.

Optimization is difficult to achieve and even more challenging to sustain in a rapidly evolving and continuously changing business climate. Optimization and excellence are shaped by an organization's daily shop floor routines and the discipline exercised in executing those routines while applying the lean and agile principles, practices, and methods that drive evolutionary as well as revolutionary productivity improvements. Manufacturing optimization and excellence do not happen by chance, they are the result of a good roadmap and strategic plan that provides direction and focus on what matters, leading the organization on a journey filled with challenges, surprises, and hopefully, lots of success.

Figure P.2 presents the daily shop floor management roadmap which provides a framework for this book. The premise is to understand one's current operational maturity, knowing that it will take time, patience, and a coordinated effort, throughout the organization, to realize a state of control, optimization, and excellence. If you have the desire to acquire a deeper understanding of daily shop floor management, in a lean and agile environment, this book is likely to satisfy that thirst for knowledge. At the very least, I expect you will find several "golden nuggets" that will enhance your understanding of shop floor operations and broaden your perspective on what it takes to be a more effective employee, manager, or organizational leader. At any point in time, if you have questions, comments, or suggestions to improve this manuscript, please feel free to write me an email at: gisi.opexcellence@gmail.com. I hope you find this book interesting and useful, now and in the future.

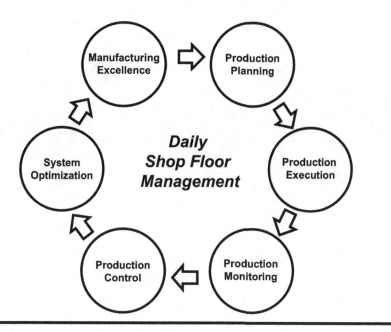

Figure P.2   Daily shop floor management roadmap.

Sources

[1] Philip J. Gisi, 2018, *"Sustaining a Culture of Process Control and Continuous Improvement: The Roadmap for Efficiency and Operational Excellence"*, New York: Taylor & Francis Group

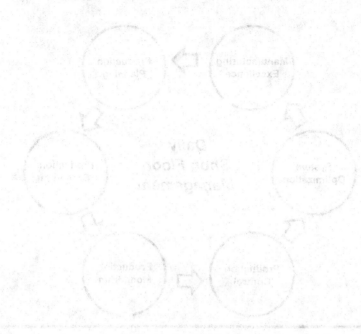

Figure 2.2. Daily shop floor management cycle.

Acknowledgment

This book is the result of years in industry and academia, working with and learning from others. My writing comes from a place of curiosity, observation, experimentation, conversation, problem-solving, consulting, and most of all, appreciation. Appreciation of the many people in my life and career, who have coached, challenged, frustrated, and guided me to where I am today. This book is just a stepping-stone along the journey of life. It's an opportunity for me to give back a little bit of what some many others have given me over the years. Thus, I share this knowledge in recognition of all those who made it possible. Although I can't possible thank everyone personally, you know who you are, and I thank you! Special thanks to my wife Kathleen and the family members who kept me grounded and sane during the writing process.

To the reader, thank you for taking interest in my book. Writing a book of this nature, or perhaps of any nature, is not easy. In fact, it's humbling. As you work to formulate your ideas, you come to realize the possibilities, the board nature of a topic, and the need to create a framework within which to scope your writing. I decided to focus on the shop floor because, to many, it's a black box of activity that spits out products, which are sold to customers, so we can make money and grow the business. In reality, it's not a black box but, for the most part, a dedicated group of unique individuals working toward a common goal. In its best form, it's a structured and disciplined operation that takes leadership, expertise, know-how, know-why, commitment, and lots of flexibility to deliver on customer expectations. Managing a shop floor is hard work and often a miracle of resilience and persistence. Regardless, the shop floor is a world unto its own, and a place few have traditionally dared to tread. However, it's time to change that mindset and make it a workplace for everyone to appreciate and embrace. Enjoy the book!

About the Author

Philip J. Gisi, PMP, SSMBB, has over 30 years of experience in the automotive, commercial, and aerospace industries. His areas of expertise include new product development, process technology, quality management, automotive electronics, and manufacturing operations. He has worked internationally in multicultural environments and is currently involved in project management and efficiency improvements as an internal consultant at Vitesco Technologies. Phil has a Master of Science in Engineering, a Master Certificate in Project Management, Lean Management Certification from the University of Michigan, Project Management Professional (PMP) Certification, and a Six Sigma Master Black Belt (SSMBB). Phil has been sharing his experiences in Project Management, Lean Concepts, and Six Sigma Process Improvement as a DePaul University instructor for over 20 years and is the author of the book *Sustaining a Culture of Process Control and Continuous Improvement: The Roadmap for Efficiency and Operational Excellence*, published by the Taylor & Francis Group.

Introduction

Overview

Thriving in today's business environment requires companies to continuously strive for operational excellence at all levels of the organization. Simply working to maintain existing operations is not an adequate or sustainable business strategy, especially when competing in a global market. To remain relevant, companies must adopt a process control and continuous improvement mentality as an integral part of their daily work activities. These two operational disciplines form the foundation and steppingstones for manufacturing excellence. Processes must be stable, capable, and controlled as a prerequisite for sustainable improvement. Sustainable improvements must be strategic, continuous, and focused on process optimization.

As companies work toward excellence, they can't lose sight of the fundamental activities necessary to preserve what they have already achieved. This is the role of **daily shop floor management (DSFM)**: to maintain operational performance while realizing and sustaining productivity improvements on the never-ending journey of operational excellence. In this regard, it's important to understand the key principles and exercise lean practices associated with well-established and managed manufacturing operations, regardless of products produced or services provided. This book is intended to provide a comprehensive look into the behaviors and work routines required for establishing a well-run, disciplined manufacturing system.

Manufacturing processes form the foundation for operational execution and provide a framework within which work is accomplished. Processes help to produce goods and provide services that generate income for profitability and growth. **Standardized processes** promote operational stability and predictable output. They serve as a baseline for process control and continuous improvement. Standards, when followed, help to reduce process variation which can be disruptive to a manufacturer's ability to achieve and maintain stable and capable processes. Once operating standards are established and followed, processes must be continuously monitored for abnormalities and defects, which should be highlighted and targeted for elimination. A defect occurs when one or more product characteristics (form, fit, or function) does not conform to requirements or services don't meet customer expectations.

Deviations from standards can be considered process abnormalities that may result in defects, if ignored. Process deviations are highlighted through visual controls and corrected through **deviation management**. Deviation management focuses on understanding the underlying cause of an undesirable event and should result in the implementation of a countermeasure intended to prevent defect reoccurrence. Occasionally, a problem may occur that needs further study to identify its root cause for elimination. This can be achieved through **structured problem-solving** for quick and efficient problem eradication.

Ignoring significant deviations can lead to product defects and erosion of process stability. Defect elimination improves line stability which is maintained through continuous process monitoring and control. **Process control** is the act of maintaining process stability and capability. **Process stability** is characterized by consistent and predictable process output over time. **Process capability** builds on stability and reflects a process that delivers products and services that consistently meet customer requirements. Control is evident when a process is continuously monitored for abnormalities, and countermeasures are taken to correct significant deviations from the norm.

Control of the shop floor requires **organizational structure, operational discipline, and employee accountability**. Organizational structure provides a framework and direction for shop floor management. Operational discipline is the ability to effectively implement the manufacturing system as defined and reflects the confidence, commitment, and belief a company has that their organizational structure, when executed properly, will yield desired results. Employee accountability reinforces the behavior expected of employees to execute their assigned manufacturing work routines, properly and completely, with clearly defined consequences for not doing so. Clarity of expectations must be provided to all employees for effective shop floor management.

Standard work instructions are often used to articulate the expectations of manufacturing operators performing time-based or cyclic work. Effective work instructions define the work activities, sequence, timing, and expected results. Expectations of shop floor workers, supervisors, support functions, and managers can be expressed and managed through **standard work routines** (SWR). Standard work routines describe the tasks and frequency of shop floor work activities that must be performed, by support personnel, to ensure process stability, capability, and control are maintained and the process is continuously improved. Activities required for shop floor management are typically outlined in a **manufacturing operations manual** (MOM) that establishes the framework for conducting daily operations. Standard work routines can be standardized to a degree but must allow for some customization and flexibility to accommodate the unique differences in organizational structure, operational dynamics, and employee expertise.

Every system, including the shop floor management system, requires a periodic "check and balance" to confirm its continued effectiveness over time. This can be accomplished through **layered process audits** (LPA). Layered process audits is a method for observing and validating process execution to ensure the manufacturing system is being properly implemented and continues to function as intended. Significant deviations identified during these audits must be highlighted, prioritized, and corrected, when necessary. Layered process audits are performed to ensure system integrity is maintained according to operating parameters and standards while product inspections confirm conformance to requirements. If the company is not continuously improving, their competitors are and will eventually lure existing customers away with higher quality, lower cost, and a more attractive product portfolio.

Continuous improvement is another vital component of a robust shop floor management system. Proactive improvement techniques such as value stream mapping and design, Kaizen events, and Jishuken workshops can be used to enhance operational performance. Value stream mapping highlights the disruptions to material and information flow throughout the value stream. These disruptions can then be systematically removed to create a more reliable and predictable process. Kaizen events target a known process improvement for realization while Jishuken workshops take a self-study approach to identifying and eliminating waste revealed through observation, analysis, and immediate action.

The methods discussed are a few of the many approaches used to manage and improve shop floor performance. In effect, shop floor management is intentional, structured, disciplined, and continuous. It takes a community of like-minded individuals to exercise their knowledge, experience, and the authority needed to ensure an efficient and effective manufacturing system is in place, followed, and continues to produce desired results. As with any system, it must be maintained through active, engaged, and committed management.

Book Contents

Excellence in daily shop floor management takes a holistic approach. This approach includes having the right culture, organizational structure, operational discipline, and employee accountability. It also requires a robust manufacturing management system that must be continuously monitored, controlled, and improved. In the following chapters, this book will take you on a journey that outlines a robust manufacturing system and the activities for process optimization and operational excellence from the perspective of shop floor management. Let's begin.

Part 1: Shop Floor Foundation

Part 1 of the book contains four chapters which outline a solid foundation for manufacturing management. The first chapter (1.1) is about **organizational culture** which touches on the mindset, commitment, and alignment of organizational needs to successfully achieve manufacturing excellence. If the right culture is not in place to establish how an organization should think and act for success, future challenges may become obstacles and expectations never realized. Chapters 1.2, 1.3, and 1.4 explore the roles of **organizational structure, operational discipline, and employee accountability**, respectively. Organizational structure provides a framework and direction for shop floor management. Operational discipline is the ability to effectively implement the manufacturing system as defined and reflects the confidence, commitment, and belief a company has that their structure, when executed properly, will yield desired results. Employee accountability reinforces the behavior expected of employees to execute their assigned work instructions and work routines properly and completely, with clearly defined consequences for not doing so. Clarity of expectations is essential for all employees. In essence, the first section of this manuscript outlines the foundational elements for daily shop floor management.

Part 2: Shop Floor Fundamentals

Part 2 elaborates on the fundamental practices of daily shop floor management. Topics include **process standardization** (Chapter 2.1) used to baseline manufacturing performance and control process variation. Chapter 2.2 touches on **data and information** gathering, analysis, and reporting used to assess and regulate performance. Chapter 2.3 stresses the importance of **data visualization** to display and monitor key process indicators for nonconformances to operating instructions and standards. This topic serves as the basis for deviation management. Chapter 2.4 goes on to cover the role of **communication** in the dissemination of data and information to appropriate functional groups and employees for timely consideration and action to ensure a robust shop floor management system is functioning adequately.

All these activities center around **people** having the right knowledge, skills, experience, and mindset to make the manufacturing system work as designed. This is the subject of Chapter 2.5. The final chapter in Part 2 focuses on the discipline of **facilities management**. Although this book concentrates on shop floor management, we need to understand all the factors that make manufacturing possible. This includes the environmental, health, and safety aspects of facilities management, including temperature and humidity control, technical cleanliness, utilities, ergonomics, and electrostatic discharge (ESD). These essentials are an inherent part of a well-managed, operationally excellent organization.

Part 3: Shop Floor Management

Part 3 dives deeper into the heart of shop floor management by looking into the elements that drive the manufacturing system. It brings together the topics under shop floor foundations and fundamentals, discussed in the previous two chapters, to create a structured and disciplined approach for effective **manufacturing system execution** (Chapter 3.1), **monitoring** (Chapter 3.2), **and control** (Chapter 3.3) of the shop floor. Management system execution focuses on a defined set of periodic tasks or standard work routines intended to direct daily work activities to ensure work priorities are completed and completed correctly. The chapters on monitoring and control outline the on-going activities needed to ensure process stability, capability, and control are achieved and maintained through the application of deviation management and problem-solving techniques.

Subsequent chapters, Chapters 3.4, 3.5, 3.6, 3.7, and 3.8, consider the contribution of people (man), equipment (machines), materials, methods, and the environment (4 M's+E) in daily shop floor activities, respectively. These key operating factors need to be carefully managed to ensure a portfolio of products are manufactured to requirements, shipped in the right quantities, delivered on time, and meet customer expectations. In essence, people need to be developed, machines need to be maintained, materials need to be properly handled, methods need to be effectively executed, and the work environment needs to be properly controlled to realize business goals and objectives.

Part 4: Shop Floor Improvement

Part 4 of this text explores the activities for continuous shop floor improvement. In this section, you will encounter practices employed for incremental improvement, process optimization, and manufacturing excellence. Topics that will be discussed are product and process development (4.1), productivity improvement (4.2), agile methods (4.3), shop floor optimization (4.4), and manufacturing excellence (4.5). Many tools and techniques will be introduced to the reader and details behind deployment of these activities are provided. Continuous improvement is an expectation of manufacturing since low-cost countries keep the pressure on high-cost countries to remain competitive in order to grow within a global marketplace.

Part 5: Next-Generation Shop Floor

The last section of the book, Part 5, contains only one chapter that explores trends in manufacturing technology and ways in which technology can improve operational efficiency and create a "work from anywhere" reality. These are the technologies shaping the future of manufacturing. The impact of automation, big data, data analytics, artificial intelligence, predictive maintenance, mobile devices, and the internet of things (IoT) will all play a significant role in future work and influence the way we think and interact with process hardware and software. Automation is being increasingly deployed to reduce the presence and intervention of humans in production. Big data will facilitate the analysis of cause-and-effect relationships that will recognize changing equipment behavior and outcomes for predictive maintenance. Artificial intelligence will accelerate learning and apply this capability to increased manufacturing accuracy, and efficiency, independent of human interaction. This, in combination with the IoT, will accelerate awareness of potential production issues leading to a more predictive and proactive response to changing manufacturing conditions.

The Journey to Manufacturing Excellence

Improving operational efficiency is not a trivial task. Improvements become increasingly more difficult as "low hanging fruits" for reducing waste become exhausted. To breathe new life into companies struggling to remain competitive, the concept of operational excellence, or in the case of this book, specifically manufacturing excellence, has slowly worked its way into business conversations as a next step in performance improvement. To date, manufacturing excellence has not been well defined or articulated, leading to different interpretations of what it is and how best to pursue it. Let's take a deeper look into manufacturing excellence relative to the shop floor and the corresponding activities required for sustainable efficiency improvements.

Manufacturing excellence is not a goal or an achievement, it's a presence of mind, a state of existence. It's not obvious, it's a feeling, a sense of calm and control, it's about knowing what to do in response to normal and abnormal circumstances encountered every day on the shop floor. It's reflected in leadership decisions, management practices, and employee behaviors. It's evident in the attitudes of people, how they interface with management, interact with each other, and feel about the company. It's observed in their sense of responsibility, their willingness to make a difference, and in the actions people take in response to uncertain or unknown situations.

On a more practical or tangible level, manufacturing excellence is built on a foundation of principles and practices, within an organization, that help people realize the attributes associated with excellence as reflected in their attitudes, work habits, and job expectations. Although these principles and practices may vary, from one enterprise to another, certain practices, especially those used to interact and manage people, should be universal.

Manufacturing excellence starts to become a reality when optimization moves beyond individual processes toward integrated ones. Process enhancements, at this stage of maturity, are realized through the disciplined removal of obstacles hindering the smooth, steady flow of material and information within a product family's value stream. This is where value stream design is employed in the service of operational excellence.

Manufacturing excellence can be described as the application of lean and agile knowledge, skills, tools, and techniques to achieve value stream optimization through incremental efficiency improvements. It's important to note that operational excellence is not an end goal, it's a continuous improvement journey. It's a holistic strategy that focuses on delivering customer value through process control and efficiency improvements leading to optimized performance. It starts with alignment of management systems with the organizational strategy and leads to an incremental change in corporate culture as strategy deployment shapes employee attitudes and behaviors to deliver desirable and consistent output performance.

The road to optimization takes time, requires a clear path and structured approach to realize substantial benefits that deliver lasting stakeholder value. The following sequence of steps provide a solid foundation for a more efficient and competitive business position when pursuing excellence with knowledge, intelligence, discipline, and commitment.

Step 1: Develop a business strategy and deployment plan; execute, monitor, and control planned activities

Step 2: Define and implement the policies, practices, systems, procedures, work instructions, and standards required for efficient and effective operational execution

Step 3: Establish process stability to realize consistent and predictable output performance

Step 4: Confirm process capability to consistently achieve output performance expectations

Step 5: Maintain a state of process stability and capability (e.g. shop floor control)

Step 6: Continuously reduce/eliminate process waste

Step 7: Optimize material and information flow throughout the value stream
Step 8: Continuously strive for manufacturing excellence

Daily Shop Floor Management

Manufacturing excellence requires a well-defined and executed management system designed to maintain process control and drive continuous, sustainable, and efficient improvements. This ability must be reflected in an organization's daily work routines and longer-term strategic projects focused on sustainable growth and improvement. In essence, manufacturing excellence relies on daily shop floor management as a core competence for sustainability.

Simply stated, daily shop floor management (DSFM) involves the activities and routines required to keep all processes in control (stable and capable) in addition to making sustainable improvements in operational efficiency. For many of us working on the shop floor or working closely with shop floor personnel, this is what we are hired to do, process control and continuous improvement! Daily shop floor management requires the coordination of activities among the production workforce and support functions to produce consistent quality parts on time and in the quantities requested by customers. Quality parts are those that meet product requirements and satisfy customer expectations.

Successful daily shop floor management requires a robust organizational structure, persistent operational discipline, and holding employees accountable for properly completing their daily work routines. Shop floor management focuses on daily operational performance and is defined by the activities completed, on a periodic basis, to maintain shop floor stability and drive continuous improvements in safety, quality, productivity, and costs. If properly planned and executed, DSFM will help ensure the manufacturing system is working as intended and delivering desired results. The corresponding sub-systems should highlight process abnormalities or problems so that appropriate and timely countermeasures can be taken to address or prevent disruptions in expected output performance. To be most effective, DSFM must identify and eliminate problems to prevent their reoccurrence.

Daily shop floor management does not stop with maintaining a stable and capable process resulting from process control. **Productivity improvement, process optimization, and operational excellence** must be part of the organizational strategy and shop floor management activities. Time must be allocated to work on strategic projects, develop employees, and experiment one's way to a higher level of performance while not losing focus on maintaining what has already been achieved. Continuous improvement is an organizational discipline and survival strategy to combat the threat of global competition while stimulating growth. Productivity improvement requires continuous employee engagement as companies work to remain "one step ahead" of their competitors. Evolutionary as well as revolutionary or "breakthrough" improvements must become the "new" norm, if companies expect to realize their ambition for survival, growth, and excellence.

Daily shop floor management is the responsibility of managers, supervisors, and team leaders who must direct, coach, and mentor employees to deliver profitable, quality products on time. It must be supported by defined roles, responsibilities, and competencies for process control and improvement. Effective DSFM is an intentional, structured, and disciplined activity. It can be complex but the objective is to make the operating system as easy as possible to understand, execute, maintain, and improve. This is not necessarily a simple task. Manufacturing value is often realized when simplicity is achieved.

This book will serve the reader by providing value-added methods, tools, and techniques to help you plan, execute, monitor, and control your shop floor management system every day, in pursuit of manufacturing excellence. It will not be easy but most things worth achieving are not. If done with commitment, structure, discipline, and accountability thrown in, it can be a very rewarding journey. Before moving forward, let's take a minute to review some of the essential practices of shop floor management. These practices will be explored in a lot more detail throughout the book and serve as the basis for a high-integrity system.

Essential Practices of Shop Floor Management

Before leaving the book's introductory chapter, it's important to outline some of the essential practices every organization should consider when moving toward a leaner operating system with excellence in mind. One objective of this text is to provide actionable practices continuously throughout the book. The following is an example of what I mean. This list provides an overview of the "essential" elements for managing daily shop floor activities. We will elaborate on each of these topics throughout the book. Some excellent practices to consider now are as follows:

Manufacturing operations manual – much like a quality manual, a manufacturing manual defines the shop floor operating system in detail. It provides a framework of actions and a roadmap of activities for manufacturing control, continuous improvement, and optimization. It defines the meaning of excellence for a company and offers a vision and direction on how to get there. As a living document, it should be periodically reviewed and updated with lessons learned and best practices.

Standardize work processes – establish operating standards to reduce process variation and create a baseline for manufacturing control and improvement. Standards serve as a reference for deviation management and layered process audits, both of which are essential practices for sustainable shop floor management.

Visual controls – good visual controls make it easy and quick to identify process deviations and allow organizations to effectively communicate and manage real-time operational performance. It's important that visuals include significant process inputs and activities so that unacceptable deviations can be identified and corrected before they can negatively impact output results.

Standard work routines – standard work routines detail how the operating system is deployed, maintained, and improved. It describes the who, what, when, where, how, and how often operational activities are to be performed by shop floor personnel to ensure a robust and effective manufacturing system is being executed.

Layered process audits (LPAs) – LPAs are used to confirm the shop floor management system is being followed and generating the intended results. It's a way to monitor and control the operating system and triggers actions to make course corrections when necessary. LPAs help to avoid system degradation over time.

Deviation management – this is a product and process control methodology used to identify and eliminate deviations from operating standards and targets. Significant deviations must be identified and quickly eliminated before they negatively impact manufacturing performance. Systems quickly deteriorate when deviations are not continuously addressed.

Structured problem-solving – a methodology-based problem-solving discipline can be used to drive efficient root cause identification for permanent defect elimination. This helps prevent the reoccurrence of defects and incrementally eliminate waste.

Shop floor walks – sometimes called lean, management, or Gemba walks, these regular visits to the manufacturing floor increase management awareness of what's actually happening and allows them to encourage and coach employees in acceptable workforce practices and behaviors.

Process control – the activity of maintaining process stability and capability for consistent and predictable output performance is an expectation of all manufacturing operations and the foundation for sustainable improvements.

Continuous improvement – continuous manufacturing improvements are essential for safety, quality, productivity, and costs. It's a survival tactic for those in a competitive industry and essential for the pursuit of manufacturing excellence.

Motivate, empower, develop, and recognize employees – happy people are productive people. Shop floor management requires attentiveness to equipment, materials, methods, and the shop floor environment, as well as people to achieve excellence.

Preventive maintenance – there is nothing much worse in manufacturing than to have a piece of equipment go down when it's needed most. To avoid this situation, preventive maintenance schedules should be prepared and executed as planned. Just like people need food and water to function, equipment need periodic maintenance to continue working as designed. If you neglect this responsibility, don't expect equipment to function indefinitely.

Strategy deployment – a properly deployed strategy communicates a vision, mission, and expectations for the future and helps align organizational personnel to work in concert toward common goals and objectives. In the absence of clear direction, operational efficiency is often lost as people individually decide what's important versus the collective wisdom of leadership.

Communicate – people need timely data and information to complete their daily work routines efficiently and effectively. These data and information need to be accurate, factual, and timely. Poor communication can be a source of friction and conflict, leading to operational inefficiencies. People work best when they are aware of what's needed to complete the work for which they are responsible.

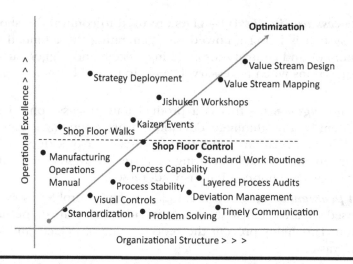

Figure I.1 Principles and practices of daily shop floor management.

Lessons learned – capturing lessons learned is not something nice to do, it's an operational excellence imperative. Many continuous improvements are likely to come from the knowledge and experience gained from what's gone right and wrong on the shop floor. If organizations are not learning from these experiences, they are losing ground to their competitors who are likely doing so.

Prioritize, prioritize, prioritize – since time and resources are limited, it's important to continuously focus on what matters. There is a lot of wasted time and effort in industry. Know what's important for the business and prioritize your activities accordingly. If an activity does not contribute to the goals, objectives, or bottom line of the organization, consider not doing it or stopping it. Question the value of every task you do and only do those that make sense. Prioritize!

Many of the practices presented are based on a set of principles which serve as the foundation for daily shop floor management. The principles behind manufacturing excellence can vary from one organization to the next. However, the principles presented above are fundamental and "core" to mostly all manufacturing facilities seeking operational excellence. These principles have been visualized in Figure I.1 and will be discussed in more detail, with others, in subsequent pages of this text.

SIDEBAR: EIGHT STEPS FOR IMPROVING SHOP FLOOR PERFORMANCE

■ **Exercise the 5S methodology.** A clean, organized shop floor boost employee moral and creates a more efficient work environment

- **Standardize your processes.** No standards equal no baseline for process control or improvement.
- **Implement visual controls.** Make data and information visible. Easily understood data and information facilitates process monitoring and control.
- **Practice Deviation Management.** Deviation management is essential for maintaining process stability, capability and control to sustain manufacturing expectations.
- **Conduct Layered Process Audits.** Audits allow you to assess system performance and ensure the system is working as intended and producing desirable results.
- **Daily Shop Floor Walks.** Walking the shop floor heightens awareness of production activities, problems and opportunities for improvement.
- **Engage in Problem Solving.** Continuously identify, prioritize, analyze, and implement root cause corrective actions to eliminate problems from reoccurring.
- **Exercise Discipline and Accountability.** Reinforce the organizational structure and operational discipline to follow the established process and hold employees accountable when evidence indicates they are falling short of expectations.

Summary

Exhibiting the right mindset and behaviors is the starting point for operational excellence. Standardization is the basis for process control and continuous improvement. Visual controls provide a window into operational performance and help trigger action for process control and improvement. Good communication up and down the organizational hierarchy helps maintain awareness and often triggers the need for action. Deviations from operating standards or targets must be monitored and controlled to ensure predictable outputs. Change management must also become part of the operating system to ensure sustainable process improvements.

Consistent, predictable operating performance is a key attribute of operational excellence. This is why, preventive maintenance, layered process audits and structured problem-solving are essential activities for robust shop floor management. Unfortunately, effective shop floor management can't occur without good, data-based decision-making, which is reinforced by reliable data and factual information.

Operational excellence takes time and requires a clear strategy and structured approach to realize the substantial benefits of delivering continuous and lasting customer value. It's built upon a solid foundation and framework where it can take root, evolve, and thrive when pursued with commitment, knowledge, intelligence, discipline, and accountability.

Key Points

- To achieve excellence, the company mindset must move beyond optimization of discrete work areas to optimization of product value streams that span multiple departments and functions.
- Sustainable improvements are built on employee understanding, agreement, and commitment to follow process standards.
- Management assures the shop floor management structure is set up and effectively working with employee responsibilities and accountabilities clearly defined.
- Information flow from production line employees to top-level management should happen as quick as possible.
- Perfection is an ideal, not a condition; it's unattainable.
- The absence of process ownership will result in the degradation of process integrity, leading to underperformance and operational inefficiency.
- The primary objective of daily shop floor management is to minimize process variation!
- Whatever you do, keep processes as simple as possible and challenge people to identify opportunities for improvement by asking "Can it be made simpler?"
- When properly executed, daily shop floor management will lead to a more stable, profitable, and competitive state of existence.
- The ultimate objective for a lean organization is operational excellence. This is a state of existence that is perpetually evolving, as the organization continues to monitor and adjust to new information, changing conditions, and a dynamic business environment.
- Tribal knowledge can be used to manage a good company but factual information and reliable data are required to run a great company.
- Daily shop floor management requires process standardization from which process stability, capability, and control can be preserved.
- Daily shop floor management is like an engine control unit in a car. If it's not continuously working to optimize performance, efficiency is lost and operational excellence can't be realized.
- Process control is the foundation for sustainable efficiency improvements.

A Note about Supply Chain Management (SCM)

Before starting a discussion on shop floor management, there are several key topics that are worth reviewing due to their connection with the subject. To ensure shop floor management is discussed within the appropriate context, the following pages will overview supply chain management (SCM) and three key functions within this discipline, including administration, production, and distribution. Although administration and distribution are outside the scope of this book, we will review them briefly here and in the following chapters since they play an important role in supporting the efforts that make shop floor activities possible. See Figure I.2.

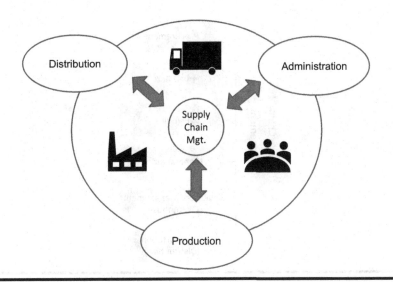

Figure I.2 Key supply chain functions.

Manufacturing Supply Chain Management

Successful shop floor management is the result of many different departments and functions working in concert to support upstream and downstream manufacturing operations. In manufacturing, this is considered the supply chain. A supply chain is a connected network of coordinated organizations or functions executing activities designed to source, produce, and transport goods for consumption, distribution, or sale. It's the act of creating a product from raw material to finished goods for transport and delivery. Supply chain management is the harmonization of logistics within the supply chain. The Council of Supply Management Professionals defines supply chain management (SCM) as encompassing "the planning and management of all activities involved in sourcing and procurement, conversion, and all logistics management activities. Importantly, it also includes coordination and collaboration with channel partners, which can be suppliers, intermediaries, third party service providers, and customers. In essence, supply chain management integrates supply and demand management within and across companies." [1].

Manufacturing supply chain management encompasses a broad discipline of activities which includes supply chain planning or strategy, raw material sourcing, manufacturing, distribution (delivery and logistics), and product returns (defective or unwanted). Effective supply change management requires that businesses demonstrate capability and continuous improvement in all five of these operational areas. The challenge is creating a seamless connection between these activities and a smooth flow of information and materials throughout the value stream.

The influence of SCM may extend into marketing, sales, product design, and information technology, depending on how an organization is structured. Logistics is a sub-category of supply chain management that coordinates the efficient movement of materials through the value stream, ensuring the proper storage of goods

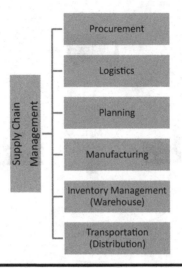

Figure I.3 Supply Chain Management (SCM) activities.

and timely flow of information to the people who need it. Success is executing an on-time process that consistently meets customer quantity demands, quality, and delivery requirements. The scope of logistics can include inbound and outbound transportation, fleet management, warehousing, materials handling, order fulfillment, inventory management, supply planning, and management of third-party logistics service providers. In certain circumstances, logistics may also include sourcing and procurement, production planning, scheduling, packaging, and customer service.

Another aspect of supply chain management is the Supply Chain Manager. Expectations of a Supply Chain Manager include minimizing material and finished goods inventory, preventing material shortages, and keeping production costs low. In short, optimize all supply chain activities. If planned and executed properly, supply chain management can increase revenues, minimize costs, and have a positive impact on company performance. See Figure I.3 for a depiction of SCM and corresponding interfaces. Let's review three other aspects of SCM in more detail, administration, production, and distribution.

Administration

Administration encompasses the management of sales and services, finances, logistics, and people. Production involves the execution of processes and procedures to achieve targeted results while distribution is responsible for the safe and timely delivery of finished goods to their destination. To do this effectively, timely communication of data and information is required in addition to routing of material and finished goods through the facility. Key administrative activities, within supply chain management, include timely information flow to production, preparing, executing, controlling and continuously improving standard operating procedures, material handling and tracking, production and human resource scheduling, managing deviations from standards or targets, and problem-solving to prevent reoccurrence.

The administrative function behind SCM is intended to define, support, and continuously improve the infrastructure necessary for effective manufacturing system deployment. These functions serve to ensure the needs of manufacturing are met so production can focus on making quality parts that meet customer expectations. Some typical administrative responsibilities include:

- **Purchasing** – obtain quotes, select suppliers, and order materials (production and non-production)
- **Human resources** – employee on-boarding, training, benefits and salary administration, and disciplinary action
- **Facilities** – building maintenance, machine shop, environmental, health and safety, security, and food services
- **Quality** – supplier quality, product testing, customer complaints, verification, qualification and compliance, and metrology
- **Logistics** – material planning, receipt, coordination, finished goods collection, storage, and delivery.
- **Finance** – budgeting, forecasting, and cost control
- **Sales and marketing** – product promotion, customers engagement, customer satisfaction, and customer retention
- **Customer service** – manage customer concerns and complaints
- **Legal** – Regulatory compliance, taxes, and contracts
- **Warehouse** – receipt of raw material and shipping of finished goods
- **Design and development** – understanding requirements and developing products that meet customer expectations

It's important to understand that the administration team works for production. They should strive to minimize the extraneous workload they place on the production team. I once observed an administrative team put so many demands for data and information on the production team that production started to deteriorate as they slowly lost focus on their primary objective of producing quality parts for timely customer delivery. The administrative office became their primary customer. If the office needs something from the product floor, they need to get it themselves and allow production to do their job of building great product that meets customer expectations. That way, the customer keeps paying the bills and office personnel get to keep their jobs.

Manufacturing administration is a service to production. They should work to make production safer, cheaper, easier, and faster without compromising quality or delivery. They must focus on listening to their customers and working to satisfy their needs. Although an important function, administration is not the focus of this book, it's the manufacturing floor. Let's briefly overview the topic of production next.

Production (e.g. Manufacturing)

Production is the act of turning raw materials into finished goods or the assembly of components into a product for consumption or sale. It's generally considered the activity of producing, creating, or manufacturing something of value. Most manufacturing operations are composed of people, machines, methods, materials, and in

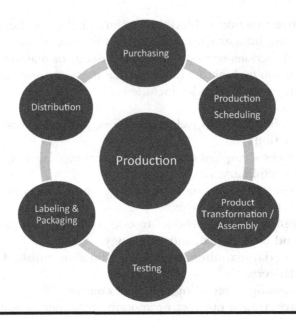

Figure I.4 Production activities.

certain cases, a controlled environment. These characteristics define the fundamental elements of a manufacturing operation. When planned, executed, monitored, and controlled properly, manufacturing systems can generate profitable goods that contribute to the overall mission and vision of a business. However, running a manufacturing operation is not a trivial task. It takes time and money, knowledge and skills, structure, and discipline to achieve a desirable, consistent, and predictable output. It takes a commitment.

People bring the knowledge, skills, and experience to the shop floor while materials, methods, and machines provide the means for transforming raw materials into finished goods. Production can also involve the assembly of components into a more value-creating deliverable. In essence, it's a holistic approach starting with the acquisition of raw material and ending with the distribution of finished goods to expecting customers. In many instances, its impact can extend well beyond the shop floor to customer support and product life cycle termination involving product disposal and recycling, from aluminum cans to automobiles. Viewed holistically, production may involve the activities outlined in Figure I.4.

High-volume production requires a skilled and coordinated team of people. Customers need to be managed, material needs to be ordered, the facility needs to be maintained, people need to be paid, and finished goods need to be delivered. To keep operations running smoothly, manufacturing is supported by teams of people working in concert to facilitate the seamless transition of raw material into finished goods. As we move forward, the shop floor will be the focus of this book since this is where value is created.

Modern-day shop floor operations leverage information technology and control systems to assist with tracking of materials, products, and work performed. In addition, testing may be required to assess the quality of production output to specifications or standards. To maintain consistent and predictable work output, significant product and process deviations must be identified, highlighted, and investigated for root cause. Appropriate actions are then taken, as needed, to maintain product and process integrity.

Inventory management is required for raw materials, non-production materials, and critical spare parts to avoid unexpectedly long equipment downtime. To maintain efficient operations, standard work instructions (SWI) and standard operating procedures (SOPs) are documented, communicated, executed, and controlled through periodic process audits to ensure consistent, disciplined operational performance. Engineering, administrative, and distribution functions are also employed to support operational activities. Let's take a moment to overview distribution activities.

Distribution

While manufacturing works to transform materials and assemble components into finished goods, these outputs must find their way to the customers, sellers, and end users of the product. Thus, another aspect of the manufacturing process, within the supply chain, is the handling of finished goods for storage and delivery. Upon completion, finished goods must be properly labeled and packaged for transport to their designated locations. The distribution function handles the logistics of getting the product to those who want or need it. In certain situations, distributors are used to handle bulk orders by purchasing product from the manufacturer and then selling smaller quantities to retail stores or directly to customers. Distribution management refers to the process of coordinating the movement of goods from the supplier or manufacturer to the point of sale. A well-managed shop floor can achieve and improve operational efficiency through strong support from administrative, production, and distribution functions. Good cooperation between these support groups will lead to stable and reliable output performance and customer satisfaction.

Source

[1] https://cscmp.org/CSCMP/Educate/SCM_Definitions_and_Glossary_of_Terms.aspx

SHOP FLOOR FOUNDATION

The shop floor system is only as strong as the foundation upon which it's built.

Objective: Understanding the key elements for planning a robust shop floor management system.

Overview

Every organization is built on a cultural foundation that defines its character. The shop floor is no different. In Part 1 of this book, we will discuss the foundational elements that influence shop floor management actions and employee behaviors. We will explore the impact of these elements on the organization to understand their effect on the work environment and employee activities.

Part 1 commences with an understanding of **Organizational Culture** driving employee behavior, its influence on shop floor management, and importance in achieving manufacturing excellence. People's attitude about work, actions, and decisions taken during work effect and influence others to behave in a similar way. This is how company culture evolves and is reinforced. It can also be reinforced or changed by the actions (or inactions) of management in response to desirable or undesirable employee habits and interactions with each other. As part of the shop floor management foundational discussion, we will touch on the topics of culture and behavior, the need for organizational structure, the role of operational discipline, and the importance of employee accountability in creating a solid foundation upon which to manage the shop floor on a daily basis.

Chapter 1.2 looks to define or refine an enterprise's **Organizational Structure** to achieve more robust and consistent product and service outcomes. It requires periodically reviewing and updating the strategy, policies, practices, and procedures the organization depends on to run the business efficiently and effectively. It's about

DOI: 10.4324/b23307-1

implementing standards and other key practices to minimize process variation and increase predictable outputs. It requires defining a workable system that is documented and executed to achieve expectations.

Once the appropriate practices and procedures are in place, focus turns to realizing more stable and predictable processes through defect identification and permanent elimination. Defect elimination leads to increased process stability paving the way for process capability. This may require further variation reduction if output performance does not consistently meet targeted requirements. Stable and capable processes and manufacturing systems establish the foundation for continuous and sustainable improvements as the journey toward process optimization continues. Once organizational structure is in place, we turn our attention to **Operational Discipline** by defining the periodic work routines (daily, weekly, monthly…) required to maintain process control rooted in standardized, stable, and capable processes. This requires the execution, monitoring, and control of daily operational activities (or routines) to sustain consistent output performance.

Daily management of the business is required by all departments at all levels of the organization. This is done by assigning work routines to key roles and responsibilities that help maintain process control and drive efficiency improvements. **Employee Accountability** for executing work instructions and performing daily work routines can be confirmed through layered process audits, periodic performance reviews, production results, performance trends, and strategic project outcomes. These activities help reinforce the right behaviors and essential actions for maintaining a transparent and effective management system.

Implementing an organizational structure, maintaining operational discipline, and ensuring employee accountability are not trivial activities. They require time to implement and must be continuously managed upon realization. There is no silver bullet in doing so, with trial and error a common practice for learning and improving each unique operating environment.

Organizations must take time to understand the expectations of manufacturing excellence at each milestone of their journey, educating and holding people accountable for short- and long-term activities and deliverables while maintaining sustainable

Figure 1.0.1   Fundamental Elements of Daily Shop Floor Management.

improvements. Organizations don't change, people change. Thus, developing people skills and the capability to change is essential for identifying and eliminating the gaps between the current and desired state of operation. The objective of Part 1 is to help organizations understand the fundamentals of shop floor management that lead them to a state of operational excellence. In the following chapters, we will review these shop floor management fundamentals as depicted in Figure 1.0.1.

1.1

Organizational Culture

Culture is what motivates and retains talented employees.

~ Betty Thompson

Objective: Plan, practice, and perfect a culture of manufacturing optimization and excellence!

Introduction

Company culture shapes how people think, feel, and act within the context of their daily work environment. Employee attitudes and behaviors are directly influenced by what they observe in leadership, their direct management, and more importantly people within their work groups. Most people mimic the attitudes, behaviors, and actions of those they serve and that of their peers to "fit-in" and be considered a "good employee" or "team player". Therefore, a culture of manufacturing excellence starts with leadership displaying and living the values they want and expect others to espouse.

Organizational values impact the fundamental beliefs exhibited in an organization's behavior and decision-making style. Values drive actions and must be considered when looking to change the way an organization thinks and acts. Values will drive how a company treats its employees. Companies that value individual talent, as a business asset, will act differently than those who see people simply as a tool to achieve their profit objectives. This will be reflected in the behavior of managers interacting with employees. To some, the values of an organization are considered an influential and enduring determinant of sustainable performance. Values are the basis from which company policies and strategies evolve. Every company and facility has a unique culture reflective of its people and management team. Manufacturing excellence starts by defining or redefining what a culture should be in terms of the values, behaviors, and norms expected of employees.

Culture = Values + Attitude + Behaviors

DOI: 10.4324/b23307-2

The ability to establish and maintain a mindset of operational excellence is rooted in an organization's values, behaviors, structure, and discipline. It requires respecting, mentoring, coaching, and continuously developing people to be the best they can be as reflected in their work routines (personal and organizational) and accountability in meeting their work commitments.

Lean Mindset and Culture

Striving for leaner operations in a manufacturing environment has its challenges. The application of lean concepts to drive efficiency improvements often involves a change in the way we think, behave, and act to produce consistent and desirable results at the lowest effort and cost possible. Creating a lean culture must start with a clear commitment for change and a vision of what that change looks like upon realization. Unfortunately, change is not easy to accomplish or likely to occur unless a meaningful commitment to change is articulated through a strategic plan: a plan developed and executed with adequate resources, management commitment, and the discipline to effectively deploy and manage the activities required to realize a future state.

A lean mindset strives to create an atmosphere of continuous improvement with a never-ending focus on the elimination of waste and obstacles that disrupt the movement of information and materials throughout the value stream. The objective is to design and continuously redesign work processes to achieve ever-greater, value-added flow by minimizing equipment idle time and people wait time. The key is to establish a process flow that rapidly and seamlessly moves material and information through production. A lean mindset also involves creating a culture that identifies, highlights, and addresses problems quickly through efficient communication and rapid problem-solving. Quality will improve with increasing process speed as products make their way to the customer on time and in the right quantities ordered.

As discussed, implementing a lean culture, in pursuit of operational excellence, often requires a cultural change. A change in operational principles, practices, and behaviors takes an action plan and time to execute it. According to an article by John Shook [1], managers can start changing their organization's culture by:

- Changing what people do rather than how they think;
- Acting their way to a new way of thinking than to think their way to a new way of acting;
- Giving employees the means by which they can successfully do their jobs;
- Recognizing the way problems are treated reflects your corporate culture.

A good plan is, a SMART plan, derived from the organizational strategy. SMART is an acronym for Specific, Measurable, Achievable, Realistic, and Timely. If you want more efficient and excellent manufacturing operations, prepare and execute a plan that will change what people do and how they do it, which will eventually change the way they think.

A culture change, reflected in human behavior and habits, takes time lots of time. It starts with a clear expression of the company's vision and mission, articulated in its values and principles, practiced in employee daily work routines and demonstrated in performance results. Lean penetrates deep and broad within an organization's DNA when the right culture and mindset are practiced over and over again.

Modern-day shop floor management requires a continuous improvement culture where every employee is looking for opportunities to make positive changes to process efficiency and output performance. Improvements come from problem-solving, lessons learned, and continuously following standard work instructions. Improvements must be integrated into employee standard work routines and operator work instructions to ensure their benefits are realized and sustained.

Cultures will evolve over time as it adapts to a changing work environment. This requires that all employees embrace change and adjust their behavior to accept approved changes by making them an inherent part of their daily work. As ideas, better practices, and lessons learned are proposed and approved for integration into operating standards, employees should be recognized and rewarded for their contribution, reinforcing the desired behaviors expected of the workforce. Clear improvement goals, with appropriate rewards and recognition, will help drive meaningful change with sustainable impact.

Key Points – Culture

■ Culture is what we see and observe in our daily work environment.
■ Culture exists everywhere. We all coexist in a world with a multitude of cultures.
■ Creating culture is easy, changing culture is hard!
■ Culture often requires cultivation to achieve a certain expectation.
■ High-volume manufacturing requires a culture where people follow rules and work to standards.
■ Preparing an organization for change requires much thought and preparation.
■ Ambiguity breeds a "wait and see" attitude. Be clear and concise about what's expected of employees.
■ We are influenced by the actions and behaviors of the people around us.
■ Respect, encourage, develop, coach, and challenge people.

Organizational Commitment

Creating, changing, or sustaining a culture requires organizational commitment. Commitment is defined as dedication to a cause or belief: dedication that is displayed through people's words and actions. It strengthens an organization's reserve and sets an example for others to follow. A shared commitment stimulates cooperation, fosters camaraderie, and builds trust. Changing an organizational commitment may require patience while people take the time needed to understand expectations and adjust their behaviors to new or changing norms. Through patience and support, people and their commitment to supporting an organization's vision and mission will grow over time. It's accelerated when people feel safe and comfortable working together, are encouraged to make joint decisions, team up to solve conflicts and overcome obstacles to successfully achieve common goals. Commitment is complemented by clearly established company policies, principles, robust operating systems, and supportive leadership.

Achieving a sustainable commitment requires a work environment where people are appreciated and respected by management and each another. An environment where individuals can challenge the status quo, freely build strong working relationships, and have fun while doing so. An atmosphere where associates can comfortably learn from their mistakes, each other, and experience setbacks without reprimand. The opposite is also true. Commitment can be quickly lost when not continuously demonstrated by leadership or when transparency, good communication, trusting relationships, and constructive interactions are no longer valued or encouraged.

Commitment comes with a price. People who make a commitment expect something in return. They expect to gain something meaningful from their involvement. Perhaps, it's an opportunity to exercise their expertise, participate in a team, develop a desired skill, learn something new, meet new people, work on an interesting problem, or demonstrate their ability to lead. When garnering commitment, clearly communicate the potential value it can offer.

When seeking commitment, start the conversation and create awareness. Clearly communicate what a person's commitment means and how it might personally impact them. Link a commitment's importance to the business strategy. Welcome people's participation and input. Encourage employees to get involved and help shape the future. Invite leadership to talk about what will be happening and its impact on the company's future. Encourage leaders to talk openly about their commitment and how it affects or impacts their way of life. A leader's commitment should be reflected in their work attitude, daily behavior, and ongoing actions. A visual display of commitment goes a long way to convincing others of its importance. People are watching and will follow leadership's example. When commitment is visual and prominent, it's more contagious.

Commitment should not be a burden for people to bear. It must strike a balance between work, recognition, and reward. Reinforce people's commitment by giving them something to do that supports their commitment and advances the company's strategic objectives. People want to feel important and valued. Give them an opportunity to contribute. If possible, align their interests with work that needs to be done. Determine how employees can provide value toward the cause. Look for ways to bring new people quickly into the fold by integrating them into teams already engaged in driving change.

There are likely a multitude of challenges and learning opportunities available for those willing to embrace them. People need to become involved in helping to facilitate change. They need to feel valued as individuals, which can be achieved by assigning them tasks and projects that can be successfully completed. This will allow them to feel good about themselves, get excited about their work, and contribute to the organization. When realizing a commitment, continuously provide support, coaching, and encouragement. Talk with people to understand their concerns and viewpoints. Encourage people to experiment as they encounter the unknown. Invest time in people to show you care about them and their personal development. This will demonstrate respect, show appreciation, support a connected workforce, and help reinforce their commitment.

Create an atmosphere and culture where people listen and respect the inputs and opinions of others while deciding on the best course of action for the company. There is no substitute for treating people well. Respect and appreciate the work people are doing; few people will be disappointed by the attention. Everyone has an important role to play in an organization's commitment to change. Therefore, it's important to stress that one should treat others as they would want to be treated.

Listening is a powerful tool and can provide insight for gauging commitment. It can garner trust and show respect for people as you focus on what they have to say. Listening can be a conduit for validating and developing people's thought process. This builds confidence and helps individuals to think more clearly, logically, and creatively. To maintain individual, group, or organizational commitment, consider giving interested, qualified, and capable people leadership responsibility. Mentor them when the opportunity is accepted. Allow them to take ownership of assigned tasks or projects and hold them accountable for clearly articulated deliverables.

Leadership commitment is one of the most important factors for leading successful and sustainable organizational change. A commitment, that is reinforced by a strategic plan, understood by all employees, and deployed at all levels of an organization, is more likely to be accepted and realized. Leadership commitment must be evident to everyone and articulated in the organizational strategy which must be deployed throughout the enterprise to ensure employee activities and decisions align with the common interests and direction to realize a collective vision.

Key Points – Commitment

- A commitment is a promise to do something; freedom to act without conditions from others.
- Commitment often requires a willingness to change. Prepare for it.
- Commitment takes time. Communicate, engage, coach, and be patient with people.
- Buy-in is achieved when agreement is acknowledged and commitment is demonstrated.
- You can influence people's behaviors, but you can't forcibly change the way people think and feel.
- Integrity is about doing the right thing even when doing so may be uncomfortable.
- You can't fix a problem people don't want to fix, and you can't change a culture people don't want to change.
- Success requires leadership's commitment to the cause.
- Leadership must articulate, harmonize, and deploy the vision, mission, and values of the enterprise to all employees, while realizing these attributes in their daily work activities.
- A successful business strategy is supported by management commitment and deployed throughout the organization by motivated and empowered employees who share a unified vision and mission while striving to achieve common goals and objectives.

Organizational Alignment

Organizational alignment happens when a company's vision, mission, and strategy are aligned with their policies, practices, and procedures. It also requires that every individual understand their role, responsibilities, and how their daily work activities contribute to overall company success. This is key to achieving operational efficiency in any organization, especially in a manufacturing environment, where alignment with the strategy is paramount.

Alignment of an organization's strategy to key business objectives, such as lean and agile practices, is essential for successful deployment. These practices must focus on the most impactful shop floor tasks to exploit their potential for improving everyone's work efficiency. An aligned and disciplined workforce is a productive and efficient workforce.

Policy deployment is a method to deploy the company's approved strategy, at all organizational levels. Comprehensive deployment helps increase awareness and garners employee commitment on a set of common goals. Through good communication and a clear understanding of expectations, the company-wide strategy can be transformed into long-term strategic goals and actionable tasks for execution, monitoring, and control. Encourage cross-functional cooperation by promoting next-level discussions to help the organizational focus on shared goals that have the potential for achieving breakthrough results. To ensure success, actions and targets should be periodically reviewed and obstacles removed to maintain continuity and promote progress in executing the company strategy. This topic will be discussed in more detail later in this text.

SIDEBAR: A MULTICULTURAL WORLD

According to the Merriam-Webster dictionary, culture is defined as:

- "The customary beliefs, social norms, and material traits of a racial, religious, or social group; the characteristic features of everyday existence (such as diversions or a way of life) shared by people in a place or time".
- The set of shared attitudes, values, goals, and practices that characterizes an institution or organization.
- The set of values, conventions, or social practices associated with a particular field, activity, or societal characteristic.

As you can see from these various definitions, culture exists in many forms and in many facets of our lives. It exists within our family, among friends, in our social groups, and in the workplace. Often, to feel comfortable or "fit in" in a group or organization, we may adjust our behaviors and sometimes attitudes to assimilate.

When people are hired into an organization, they are expected to accept a new culture and may even be required to alter their behavior, however subtle. Leadership "defines" company culture through their behaviors and actions (or inactions), not their words. Management becomes the conduit through which organizational culture is defined, reinforced, and amplified throughout the hierarchy, for better or worse.

Throughout our lives, we pass from one culture to another as we move between our home, work, social circles, and community engagements. In doing so, we adjust our behaviors to assimilate ourselves into our current environment, allowing us to be accepted, participate, and often thrive within the shifting norms of our daily lives.

SIDEBAR: INFLUENCING COMMITMENT

Understanding what influences organizational commitment will help strengthen that commitment. Some of the organizational characteristics that can influence an individual's commitment to a group or company include the following:

- Employee engagement.
- Leadership behaviors.
- Organizational pride.
- Workplace suitability (e.g. atmosphere, cleanliness).
- Team cooperation.
- Transparency and accountability.
- Documented policies and procedures.
- Enterprise practices.
- Good communication.
- Recognition and reward.
- Shared values.
- A culture of trust and respect for the individual.

Source

[1] Shook, J. (2010). How to change a culture: Lessons from NUMMI. *MIT Sloan Management Review, 51*(2), 63–68.

1.2

Organizational Structure

Every company has two organizational structures: The formal one is written on the charts; the other is the everyday relationships of the men and women in the organization.

~ Harold S. Geneen

Objective: Preparing an organizational framework and manufacturing management system that meets business needs and achieves customer expectations!

Overview

Effective business management relies on a high-performing organization rooted in a robust management system. This system often starts with a clear strategy and management plan on how an organization wants to run its business. A business strategy defines the long-term objectives of an organization, communicated through its

DOI: 10.4324/b23307-3

vision, mission, and goals. Methods and procedures establish the tactical deployment of a management system to conduct on-going business operations. The tactical approach defines the principles, practices, systems, standards, and work instructions necessary to perform the daily activities required to maintain process stability, capability, and control.

The tactical part of organizational structure establishes the boundaries within which employees are expected to work. It defines the operating system companies use to manage and improve operations. Structure brings definition, direction, order, discipline, and accountability to an organization. These characteristics are observed in the sub-systems, procedures, methods, and tools deployed by the association. Direction is established by enterprise strategy and policy deployment. Order is evident in the degree of employee accountability exercised by the company. Operational discipline is reflected in the frequency at which supervisors and managers confirm compliance to operating standards and systems, and their response to significant deviations from processes and procedures. In short, organizational structure sets the stage for how efficiently and effectively employees work together within the system to achieve the goals (outcomes) and objectives (specific actions and measurable steps) of an organization.

The opposite of structure is chaos, a state of unpredictable behavior. The absence of structure allows employees to influence the process and others at will, offering little to no focus or direction. Chaos favors strong individuals, those likely to influence and drive organizational outcomes for their benefit, at the expense of the greater enterprise. In essence, a lack of structure is a formula for inefficiency and underperformance.

Organizational structure should define the fundamental elements of an organization's management system and growth potential. Modern-day businesses need an organizational structure that facilitates a fast, efficient, and flexible work environment for competitive advantage. Operational efficiency, the product of a well-defined and managed structure, results from standardized, stable, and capable processes that serve as a foundation for sustainable improvements required for process optimization and manufacturing excellence.

Organizational Structure = Strategy + Operating Standards + Methods + Processes + Procedures + Tools + Techniques

Organizational structure should outline the fundamental framework within which management and employees will work to achieve predictable and sustainable results. These results are defined by the short- and long-term goals and strategic objectives of the organization. The management system should embody this framework. It's characterized by business policies, realized through the execution of documented procedures, and sustained through daily work routines to ensure compliance and continuous improvement of the operating system.

By understanding the foundational elements of an operating system, we can continuously evaluate its weaknesses and gaps in structure and work to strengthen the building blocks required for maintaining process control and driving sustainable efficiency improvements: two hallmarks of manufacturing effectiveness and excellence. In this chapter, we will explore the fundamental elements of organizational structure

Figure 1.2.1   Elements of organizational structure.

and work to create an understanding of their importance in achieving operational excellence. Key elements of organizational structure are shown in Figure 1.2.1.

Organizational Framework

Every organization needs to articulate a purpose, so employees know how to orient themselves and behave within their work environment. A lack of purpose or direction can create confusion as to an organization's identity and mission. Identity and mission are important since they will likely influence an organization's culture and decision-making process. Lacking a clearly defined purpose fuels inefficiency as employees struggle to understand their role and prioritize the work required to meet stakeholder expectations. Considering this, people will default to pursuing their own work priorities, without regard for the greater organization. At its worst, a lack of alignment can lead to chaos as employees and departments work independent of each other to justify their existence by "adding value" within the fog of anarchy. A well-coordinated and functional system can only exist when employee activities align with company goals and objectives. The absence of a stated and well-communicated purpose becomes a significant obstacle to manufacturing excellence.

Every manufacturing facility has an operating system. These systems may be well defined and highly functional such as a pharmaceutical company under the scrutiny of the Food and Drug Administration (FDA). The other extreme may consist of an operating framework with little to no documentation or outline of what the structure entails, perhaps in the case of a small "Mom and Pop" shop where long-term, highly experienced workers perform the same tasks each day. The larger the operation and more complex the product, the more likely an organization would benefit from a clearly articulated system that is executed with discipline and accountability.

A stable and productive manufacturing operation requires clearly communicated expectations that create a structured and disciplined work environment where every employee knows their role, responsibilities, and daily tasks, as they work to execute processes, manage deviations, solve problems, and implement sustainable improvements. A formula for success is not inherently obvious, it needs to be deliberated, calculated, cultivated, and continuously refined. It needs to be planned, documented, communicated, understood, accepted, executed, monitored, controlled, and improved. Success is likely to be illusive, if not intentional. Success starts with a clear tactical plan of action as to how it will be achieved one employee, one task, one day at a time. A best practice is to create a **manufacturing operations manual (MOM)** or guideline within which the actions and targets for success are evident. We will discuss the MOM in more detail momentarily. Let's touch on some other elements prior to doing so.

Business Strategy

The purpose of a business is often communicated through a company's vision and mission which can be further articulated through strategic planning and deployment. A company strategy typically outlines how an organization plans to compete in the marketplace and pursue profitable growth. It presents a roadmap for action and can be complemented with guiding principles and practices. Executing a strategy, in the context of shop floor management, and manufacturing excellence, requires a clear framework within which to work effectively. This framework can be defined by the organizational structure, executed through operational discipline, and verified through employee accountability, all of which are necessary to maintain process standards and drive continuous improvements. Organizational structure outlines the processes, procedures, methods, tools, and techniques that comprise the business management systems used to maintain daily operations while incrementally improving productivity.

An organizational structure that leads to manufacturing excellence requires alignment of its policies, procedures, systems, processes, and standards. This alignment helps to drive efficient operational execution, agility, competitiveness, and sustainable growth. Alignment starts with an understanding of how employee roles and responsibilities contribute to the overall organizational strategy. Manufacturing excellence reflects an organization in which everyone, from the shop floor operator to the CEO, shares a common vision, mission, and direction that will prepare company personnel for the challenges ahead.

An organizational structure is put in place to serve a purpose. There is no "perfect" organizational architecture. To compete in today's rapidly changing business environment, organizations must be structured or "designed" for agility based on a set of desirable attributes. One key is to always know where you are going. You may struggle to get there but the direction must be clear and commitment evident. Structural adjustments can be made along the way based on data and observations of the current and changing situation. If too much is unknown, consider experimenting your way to success. Stop working on ideas that fail to deliver and look for new opportunities to pursue success in the wake of technological advancements. Foremost, understand enterprise objectives and prepare your organization by engaging the right people and resources to create an agile and responsive work environment.

Strategic plans intended to realize the vision, mission, and objectives of the organization must have the buy-in and commitment of the entire enterprise. A plan without commitment is just a plan, much like a leader without followers is just walking. Commitment helps ensure key stakeholders are on-board and ready to provide the support needed to advance the business strategy.

Guiding Principles

Dictionary.com defines a principle as "a fundamental truth or proposition that serves as the foundation for a system of beliefs, behaviors or a line of reasoning". Principles are foundational rules that govern the consequences of our behaviors. In the case of shop floor management, rules for the way people are managed, equipment is maintained, material is handled, and the environment is controlled will influence performance and output results. Shop floor principles and corresponding actions may read like the following:

- Treat people fairly and with respect; strive to make their work easy to understand and perform without mistakes.
- Prepare a robust equipment maintenance schedule. Honor it and work to remove obstacles to executing it.
- Work with cooperative suppliers willing to improve their performance and hold them accountable for their commitments.
- Treat material in a way that prevents damage or defects. Continuously work to find better ways for handling material.
- Continuously work to optimize process inputs and activities to achieve stable, consistent, and predictable output performance.

Whatever principles leadership decides are important to the organization's identify and culture, these principles will inevitably influence company culture and drive employee behaviors. It's one thing to simply define and communicate guiding principles but it takes a tremendous effort to live those principles every day through the words and actions of company influencers. All employees must be held accountable for living espoused principles, continuously working to reflect those principles in every decision and action they take in performing their daily work routines.

Routine Practices

Practice is the repeated application or use of an idea, belief, or method to maintain or improve performance. In a high-volume manufacturing environment, exercising standard practices is a way of life. There are practices for product assembly and testing, process control and improvement, equipment maintenance, and interfacing with people. Practice also involves performing an activity repeatedly, in order to maintain a consistent outcome, improve performance, or enhance individual proficiency. Common practices are often observed in areas of safety, quality, and cleanliness (5S) to minimize variation and maintain predictability. Standard practices can also be employed to achieve specific objectives such as high reliability, agility, flexibility, and innovation.

Shop floor management practices typically include standard work instructions, structured problem-solving, deviation management, layered process audits, and preventive maintenance. These are only a few examples of the many practices facilities engage in to run their operations in a structured, disciplined, and predictable way. Practices, exercised by an organization, define who they are and contribute to the character and culture that distinguishes them from their competition. In short, an organization's success is based on the principles and practices they choose to adopt, exercise, and continuously refine in response to changing work dynamics. What principles and practices does your organization embrace? Are they the right ones? Are they sufficient? Are employees aware of them? Are they being exercised and if so, will they help you achieve your tactical and strategic objectives?

Shop Floor Management Planning

Planning is paramount when striving to achieve excellence in manufacturing. As the adage goes, if you fail to plan, you plan to fail. A good plan needs structure, discipline, and accountability without which control, improvement, optimization, and excellence become elusive. A best practice is to prepare an MOM that clearly outlines the key practices that define your operating system. This establishes an operating framework, direction, and expectation for employees working within the system. The manual should outline the required shop floor activities, visual controls, monitoring practices, and controls necessary for process stability, continuous improvement, and process optimization, in pursuit of operational excellence. The following pages will elaborate on the MOM contents for efficient and effective shop floor management.

SIDEBAR: A MANUFACTURING MANAGEMENT SYSTEM

Manufacturing improvements are most successful when a plan or standard exists to guide the organization in maintaining and achieving a higher level of performance. The plan or standard does not have to be perfect since learning organizations use their knowledge and experience to continuously improve upon them over time. In this regard, every plant should be able to articulate their manufacturing operating system, verbally or otherwise, which serves as the basis for how they operate. This description becomes the foundation or structure upon which work is performed to maintain process control and drive continuous efficiency improvements with an eye toward operational excellence.

The Shop Floor Management system provides a roadmap and description of how the manufacturing location operates, essentially clarifying the who, what, when, where, why and how work is prioritized and performed at a location. Unfortunately, such a structure is meaningless if the organization does not exhibit the discipline to execute it, learn from it, and continuously applying lessons learned to improve it.

If a clearly defined management system exists, make a point to periodically review it for adequacy and effectiveness, making adjustments when appropriate.

If a system does not exist, review your current activities and consider what drives the work being performed. Are employees focused on the right things and is the work accomplished providing expected results? Consider if the organization is driving the system or if the system is driving the organization. Answering these questions may provide insight into your organization's ability to move forward in an efficient and effective manner. What's your current state of operation?

Manufacturing Operations Manual

There are many ways to structure a manufacturing operation. One approach is to start with an MOM, considered the "bible" of the shop floor management system. A manual of this type would provide a high-level outline of the manufacturing system, defining methods, tools, and techniques employed to ensure processes are properly executed, monitored, controlled, and continuously improved. The manual would capture the fundamental activities required for daily shop floor management and convey the expectations of all manufacturing support functions. An example of an MOM can be found in Appendix 2.

The MOM can be complemented by clearly defined *roles and responsibilities* for those expected to support production activities directly or indirectly. These descriptions are often used by Human Resources for the hiring process but can also be used to outline the scope of responsibility for individuals working in manufacturing. They would be a "second-level" description of workforce activities needed to effectively run shop floor operations. In many instances, organizations stop defining job expectations after documenting organizational roles and responsibilities. This may be sufficient for some but not others that demand a higher level of engagement and operational definition. When military-style precision is necessary to ensure product consistency and predictable accuracy in meeting product specifications, a "third level" of activities can be articulated which details specific tasks and their corresponding frequency of occurrence. These have been called "leader standard work". I prefer the more generic phase of "standard work routines" (SWRs) since these activities or tasks should not be limited to leaders.

Where multiple manufacturing locations exist in a corporate setting, each site is likely to have its own unique culture and behaviors, guided by common corporate practices and procedures. In this case, the MOM can serve to deploy company standards and elaborate on those standards to create a common approach to achieving, maintaining, and improving shop floor operations. It can also be used as a guide for preparing procedures, managing materials, maintaining equipment, developing people, and controlling the shop floor environment as well as anything else that needs structure and discipline to properly control operations.

The MOM can specify the methods, tools, and techniques used in creating a standardized approach for managing shop floor operations. The manual should align with the strategy, vision, and mission of the enterprise and be periodically updated to reflect the changing dynamics of the industry and competition. Not unlike a recipe for your favorite cake or dinner meal, the MOM should be structured in a way that deploying it will deliver desired results. It should also be understood that a document of this nature must be continuously updated to reflect best practices and lessons learned.

A good operations manual articulates how your employees are expected to manage the shop floor. It can be viewed as a manufacturing employee handbook, outlining the company mission, policies, practices, and culture, in addition to describing how work is to be accomplished. New employees will become productive, more quickly, knowing a manufacturing manual is in place to guide them. Having a manual means one employee's work will not be hindered by the absence of another. Employees can learn how to do the work required of other functions, if interested or necessary.

An MOM is the first step toward manufacturing standardization. It can be used as a reference for onboarding new employees and ensuring your team can reliably and efficiently execute their tasks with consistency. It can also help to keep critical knowledge from walking out the door when an employee leaves the company. Regardless, it must be kept up to date to remain relevant and effective.

Guidelines, Procedures, and Standards

Businesses are often composed of several core functions that make up the foundation of their operations. These may include sales and marketing, research and development, manufacturing, and service. These core functions are defined through various documents such as guidelines, procedures, and work instructions. A guideline can be considered a general rule, good practice, or recommended method. They typically reflect well-established knowledge within an organization or industry. In general, a guideline is to be followed unless valid justification indicates otherwise. Procedures capture the know-how of an organization while work instructions provide the directions for task execution.

For product-orientated businesses, manufacturing processes are at the heart of business operations. As a manufacturing business grows and processes become more complex, creating structure around daily shop floor management activities becomes increasingly more important. The ability to control and enhance operational performance, in the midst of increasing complexity, will be reflected in the organizational structure and operational discipline exercised to realize effective management practices. This is where procedures and work instructions play a key role.

Procedures and work instructions aid in regulating process variation and standardizing workflow. They reflect organizational knowledge, capture key lessons learned and company know-how. They are essential for process control by helping maintain conformance to requirements and driving operational efficiency to improve output performance. Without procedures and work instructions, chaos rains, as people struggle to determine the best way to perform their work responsibilities. The absence of key procedures and work instructions can significantly contribute to operational inefficiencies.

To maintain their effectiveness, these documents must be periodically reviewed and updated to reflect best practices and lessons learned. They are at the core of any shop floor system, serving company personnel by providing direction on how things get done, from product assembly, to process control, to quality testing, and employee disciplinary actions. In addition to capturing the know-how of an organization, they often document the know-why and provide operating system transparency.

Shop Floor System Execution

Once a manufacturing plan is completed and documented for everyone to see, understand, and reference, it's time to execute the plan. Execution of the shop floor operating system is about completing the work defined in the MOM (and other relevant documents) on a continuous basis through the coordination of people and resources while managing key stakeholders' expectations. It's about directing and managing planned work required to maintain operational stability and meeting customer product delivery timing with the right quality and quantities requested.

Shop floor system execution involves collecting information, analyzing data, making decisions, solving problems, and implementing actions. It includes executing SWRs, defined by the manufacturing team, to monitor shop floor activities and take actions to maintain stability when process deviations and abnormalities occur. SWRs articulate the daily, weekly, and monthly individual work tasks to create habits and reinforce behaviors expected of the manufacturing team. Let's discuss the execution of the shop floor management system with a deeper view of the benefits behind SWRs.

Standard Work Routines

Modern-day operating activities are demanding greater process control and discipline, as manufacturing complexity and higher precision product requirements are becoming the norm. To counteract this trend, facilities are responding with more sophisticated tooling, equipment, and processes in addition to more detailed work routines to minimize output variation and scrap. In the past, articulating work roles and responsibilities was sufficient to carry out the work needed to satisfy expectations. However, in today's highly charged work environment, more detailed procedures and work instructions are needed to realize the accuracy, precision, and predictability of deliverables throughout a product's life cycle.

High-volume manufacturing requires a series of unique but mostly repetitive work activities, performed on a periodic basis, to achieve product consistency. These repetitive work activities are important because they minimize process variation and the abnormalities teams face when navigating the daily challenges that come with shop floor management. Traditionally, these repetitive work instructions (or standard work instructions) have been prepared for production line operators exercising a high degree of repetitive work. However, this same concept can be applied to other manufacturing functions that must perform repetitive tasks as part of their on-going responsibilities. Although these work tasks are not as frequent as an operator assembly process, they do require regular execution intended to ensure shop floor stability. These periodic work tasks can occur multiple times a day, once a shift, daily, weekly, monthly, quarterly, or annually. It simply depends on the nature of the object, process, or system being controlled. Some typical examples of repetitive tasks include the following:

- Process audits (5S, Electrostatic Discharge Technical Cleanliness, Poka-Yoke, Jidoka, etc).
- Gemba walks.
- Scrap reviews.
- Performance reviews.
- Total productive maintenance (TPM).
- Standard work instruction verification.
- Statistical process control (SPC).
- Production start-up.

Maintaining operational stability requires structure which can be addressed through SWRs defined for manufacturing support functions. These routines capture the activities and frequency of occurrence that an individual or function is expected to perform on a regular basis. SWRs can be viewed as a daily, weekly, monthly, quarterly, and even annual checklist of activities that should be performed to maintain operational stability and drive incremental continuous improvements. Failure to perform these routines, at the established frequencies, can jeopardize system integrity. This can lead to process deviations and product defects, likely to stifle or destabilize manufacturing operations.

Disciplined execution of the shop floor management system needs to be clearly defined for each support function and executed with rigor. This can be achieved through the definition and deployment of SWRs for all manufacturing employees beyond the standard work instructions of line operators. Process control requires clear direction and discipline. SWRs can satisfy this need with little compromise. People may push back and claim that their work activities can't be standardized but it's surprising how often much of their activities are routine and can be organized into a set of tasks performed at regular frequencies.

The phrase "standard work routines" is preferred for management and support functions instead of standard work instructions since it distinguishes the type of work performed by employees other than line operators (Figure 1.2.2). SWRs can be considered an employee's roadmap for daily shop floor management. In essence, shop floor management is executed through the many SWRs assigned to responsible individuals in different functions and departments. Much like work instructions for assembly line operators, SWRs articulate the work expectations for supervisors, management, engineering, and administrative functions that support manufacturing. They define the key actions and frequency at which those actions are performed to maintain consistent and predictable output performance. They are often used to trigger problem-solving, prompt Kaizen events, or drive productivity improvement projects. SWRs can also contribute to changing organizational culture by helping define the actions and behaviors expected of employees.

Daily shop floor management involves executing the daily work routines and other activities necessary to keep a manufacturing facility efficient, profitable, and competitive. SWRs focus on achieving and maintaining stable, capable, and controlled processes within the shop floor domain. They can also be used to continuously

Figure 1.2.2 Operating instructions.

improve processes through incremental and breakthrough enhancements. Work routines set the stage for how people are expected to behave and act in the workplace. Unfortunately, work does not simply happen. It's best realized when planned, executed, monitored, and controlled, in alignment with the company strategy, principles, guidelines, and proven practices, providing a common and unified direction for all to follow.

SWRs are essentially a tailored checklist of activities for employees, who perform specific manufacturing functions, which are expected to be completed on a regular basis. These work routines articulate employee daily or periodic tasks required to ensure priority work is accomplished at designated frequencies to maintain manufacturing system integrity and effectiveness over time. When work routines are aligned between functions and support groups, they help ensure all aspects of the management system are being properly and completely deployed. They also provide every employee with a clear list of tasks expected to be completed daily, weekly, monthly, quarterly, and annually, to fulfill their role as essential manufacturing team contributors. When these routines are done in concert with activities such as layered process audits, deviation management, Gemba walks, and performance reviews, they can drive a powerful "closed-loop" system. An exhibit of SWRs can be found in Appendix 3.

Once the manufacturing operating system has been defined (MOM) and the actions executed by the designated employees, on-going activities must focus on maintaining operations through system monitoring and control. In the following pages, we will briefly explore the daily, yet essential, activities for monitoring, controlling, and improving a shop floor management system destined for excellence.

Shop Floor System Monitoring

Monitoring is a core activity of shop floor management. The primary objective of shop floor monitoring is to review, track, and report progress. This includes highlighting abnormalities and deviations from process standards, procedures, and performance

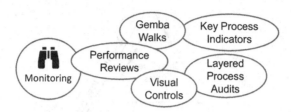

Figure 1.2.3   System Monitoring Practices.

targets for quick disposition and needed action. Monitoring also involves assessing performance data and process trends for irregularities. This requires that the right data and current information are readily available for rapid and easy interpretation. There are several practices that are exercised for effective system monitoring. They include the identification of key process indicators (KPIs), the display of visual controls, Gemba walks, performance reviews, and layered process audits (see Figure 1.2.3 for monitoring practices). Let's take some time to review these key practices in the context of shop floor monitoring and control.

Key Process Indicators

A KPI is a significant metric used to assess process performance and data trends relative to a desired result. There may be several KPIs that help identify undesirable changes to product form, fit, or function and process input/output activities. They are used to baseline performance for process control and continuous improvement and provide the data needed for problem-solving and decision-making. They are a key component of the shop floor "dashboard".

Visual Controls

A visual control is a way to draw attention to a process, product, or result especially when a problem or nonconformance occurs. A dashboard can be used to display the KPIs of an activity or area in a way that makes these metrics quick and easy to monitor for abnormalities or deviations. Rapid identification of process or product abnormalities should garner immediate attention and response, as needed. For example, a green, yellow, or red color-coding system can be a simple form of visual communication of process or equipment status. Visual controls are designed to create awareness, influence behavior and drive action.

Gemba Walks

A Gemba or shop floor walk is a form of shop floor management and employee engagement. It prompts managers and leaders to go to the factory floor and observe activities and corresponding output results at the front lines of production, where value is created. These walks help elevate awareness and enhance communication between management and the workforce through data visualization, discussing concerns and reviewing on-going operational activities. It's also a way to monitor for problems and drive immediate action to maintain shop floor system integrity.

Performance Reviews

Performance reviews are intended for management to assess performance trends in different aspects of shop floor operations. Good reviews probe deep into issues for understanding and are expected to trigger actions that prompt problem-solving and process improvement, whenever supporting data and information justify a response. These reviews should encourage open discussion and assign actions to owners with due dates for timely follow-up. Verified completion and effectiveness of actions taken is a critical component of successful performance reviews.

Layered Process Audits

If SWRs define the prioritized and repetitive daily shop floor management activities of the workforce, layered process audits are used to verify these on-going activities are being implemented properly, completely, and effectively. Layered process audits are performed by various levels of management to ensure different "eyes" or perspectives are continuously reviewing shop floor operations for efficient execution, monitoring, control, and improvement.

Process Control

A primary objective of shop floor management is to achieve and maintain process control through the execution of daily work routines and deviation management. A controlled process is stable (predictable) and capable (consistently meets customer requirements) over time. A consistent and predictable process is the result of a structured shop floor system executed with discipline and accountability. Daily management involves the execution of certain routines selected to ensure process stability, capability, and control prevail. Let's view some key routines that help maintain process control.

Deviation Management

This practice requires the shop floor management team to follow up on process abnormalities and significant deviations from process standards and targets. Deviations can result from daily process monitoring of visual controls, production meetings, Gemba walks, LPAs, performance reviews, and the like. The source of all deviations encountered must be understood and actions must be taken to bring them back in line with operating parameters, when warranted. Any actions completed should be verified as effective to avoid a repeat occurrence contributing to process waste. Effective deviation management results in permanent corrective action, preventing reoccurrence of an issue.

Action Item List

Identifying and managing open action items is an essential discipline for achieving and maintaining process control. Action items are the mechanism by which we correct and prevent reoccurrence of process deviations and can serve as a tool for driving sustainable process improvements. Since we all deal with limited time and resources, action items must be prioritized in terms of their importance and impact on process performance. If you are not continuously identifying actions to stabilize and improve performance, process integrity will erode over time, allowing defects to occur and compromise product and service quality. The identification and timely closure of key actions should be an integral part of a facility's daily work routines. Driving to due dates and confirming the effectiveness of actions taken are critical steps prior to their formal closure.

Structured Problem-Solving

Solving problems is an inherent part of shop floor management. We continuously experience problems due to issues with materials, equipment, people, process changes, and challenges to maintaining the desired operating environment. Everyone in the workforce should be familiar with structured problem-solving and practice it whenever the opportunity presents itself. In fact, problem-solving should become an integral part of every company's employee development plan. A structured approach to problem-solving is encouraged since using a proven methodology promotes efficient root cause identification for the deployment of a permanent corrective action. Problem-solving competence and efficiency is an essential component of manufacturing excellence.

Total Productive Maintenance

Continuous, uninterrupted operations are a hallmark of exceptional shop floor management since any unplanned equipment downtime disrupts production continuity and threatens timely customer deliveries. Total productive maintenance (TPM) helps to prevent unexpected equipment failure. It's a method intended to obtain the best performance from equipment by reducing downtime losses. This requires the development of workforce skills to perform a significant amount of their own equipment maintenance while freeing up the technical maintenance staff to engage in more complex equipment maintenance and repair. Employees should be encouraged to participate in TPM training that enhances their capability to accept additional work responsibilities and broadens their career opportunities within their scope of work.

Productivity Improvement

Productivity improvement does not just happen, it needs to be part of the organization's DNA. Taking a planned approach is the best way to deploy and ensure continuous sustainable improvements. This can be done by executing basic Plan-Do-Check-Act cycles, Kaizen events, Kata projects, quality circles, and targeted improvement workshops focused on achieving specific objectives such as process stability, capability, capacity, and rapid changeovers. The application of advanced technologies needs to be part of the plant strategy of improvement and managed through a series of strategic projects governed by the leadership team. More on this topic will be discussed later in this book.

In summary, organizational structure sets the stage for shop floor efficiency and success. It defines the processes, procedures, methods, standards, tools, and techniques that create a tailored framework for executing, monitoring, and controlling manufacturing operations. It's recommended that a manufacturing operations system be defined (or revised), based on industry best practices, and documented in a manufacturing manual that is communicated, exercised, and periodically updated with lessons learned. A clearly defined and documented operating system, executed with discipline and accountability, helps organizations hedge against increasing manufacturing complexity required to support the ever-increasing expectations of customers and the marketplace.

Key Points

- If you fail to plan, you plan to fail. Plan to effectively manage the shop floor.
- Each operating facility should create its own manufacturing management system which aligns with the business strategy yet reflects their unique cultural identity.
- Respect, encourage, develop, coach, mentor, and continuously challenge people to perform at their best.
- Building trust and reliability within an organization requires everyone to act with integrity. No integrity, no trust.
- Integrity is about doing the right thing, even when doing so may be uncomfortable.
- Process standardization is a prerequisite for process stability which, when achieved, must be continuously monitored and controlled.
- Detailed and comprehensive planning, consistent with the company's strategy, is essential for effective shop floor management.
- Corporations are run by people and are built upon the principles and practices others have espoused before them.
- The values of someone strongly committed to a set of guiding principles are likely to demonstrate behavior closer to the ideal.
- Gemba walks are a way for management to interact with the products, processes, and people in their organization. It's an opportunity to review visual controls, ask questions, discuss issues, make decisions, and solve problems.
- A poorly conceived operating system will generate inferior results when followed with discipline.
- By detailing the company hierarchy roles, responsibilities, and work routines, you're effectively keeping everyone informed on what's expected of them and setting the stage for holding people accountable.
- Detailed and comprehensive planning, consistent with the company's values, is essential when preparing a shop floor management structure.

SIDEBAR: SHOP FLOOR EXPECTATIONS

A good manufacturing system should be able to provide some of the following data and information upon request:

- Daily and weekly customer shipment requirements.
- Raw material availability and shortages.
- Finished goods inventory.
- Work-in-process inventory.
- Change control status.
- Operator availability and location assignments.
- Line output performance.
- Equipment downtime reasons (Pareto).
- Labor efficiency (units produced per employee).

1.3

Operational Discipline

There is no magic wand that can resolve our problems. The solution rests with our work and discipline.

~ Jose Eduardo Dos Santos

Objective: Follow your manufacturing system as designed and change it when it's no longer meeting expectations!

Overview

Organizational structure, complemented by operational discipline, is part of the building blocks for a stable, predictable, and high-performing operating environment. Operational discipline requires doing the right thing, the right way, every time. In fact, DuPont defines operational discipline (or OD) as, "...the deeply rooted dedication and commitment by every member of an organization to carry out each task the right way, every time". [1] This chapter focuses on operational discipline

DOI: 10.4324/b23307-4

which explores employee activities and behaviors necessary to ensure the policies, procedures, systems, and standards of the enterprise are implemented to achieve targeted and desirable results.

Discipline reflects the beliefs and values of an organization, reinforced by organizational leadership in their daily interaction with employees, at all hieratical levels. Since operational discipline is influenced by culture, an on-going effort is likely required to drive organizational change, if the current level of discipline is insufficient to meet existing market expectations.

In manufacturing operations, achieving excellence requires a degree of discipline that allows an organization to be the best that it can be. Unfortunately, exercising discipline is elusive unless everyone clearly understands organizational expectations and the responsibilities for which they are being held accountable. The "right way" of doing things must be inherently obvious and the concept of "every time", clearly evident, when promoting a strong working discipline. Knowing what to do, when to do it, and why one is expected to do it should be clearly communicated in the systems, processes, and procedures of the organization. This must be complemented by having coaches and mentors who can provide guidance when "doing the right thing" is not obvious and additional information or direction is required. Doing something "the right way" may not always be easy or self-evident but should be reflected in every employee's attitude toward work, as they strive to "do what's expected" and doing it the same way, every time.

Personal integrity is demonstrated when people do what's expected of them, the right way, regardless of anyone asking or watching. Knowing what people will do and how they will behave leads to more efficient and predictable outcomes. Integrity builds trust, reliability, and confidence: key characteristics of a disciplined organization. A defining attribute that distinguishes a good from a great performer is the consistency of their performance. Great athletes become great based on their desire and discipline to learn and continuously improve their performance over time. From a business perspective, operational discipline is demonstrated by commitment of employees to complete their standard work routines the same way, every time. As you may have surmised, discipline can't be blind. Although employees are expected to follow procedures as written and trained, they must be allowed to question why and speak up when they have identified a potentially better way of doing something.

Manufacturing discipline can be assessed, in part, through workplace cleanliness and orderliness, compliance to documented standards and procedures, and consistency in output performance over time. A method such as 5S is commonly used to assess, improve, and maintain workplace cleanliness and orderliness. It also can be a bellwether for gauging operational discipline based on employee compliance with the 5S standard, without being prompted.

Organizational structure is often documented through procedures and work instructions rooted in organizational knowledge, experience, lessons learned, and best practices. These documents serve as the framework within which operational discipline is exercised. Unfortunately, maintaining discipline is an on-going process and requires continuous employee engagement. Layered process audits (LPAs) can be used to periodically monitor employee compliance to standards and procedures intended to maintain system functionality and achieve targeted results. It involves independent verification of critical steps by co-workers or supervisors to confirm system compliance, integrity, and effectiveness.

Encouraging operational discipline requires a conscious effort to define expectations at different organizational levels with a clear strategy to nurture and reinforce those behaviors throughout the enterprise. Expectations for acceptable behaviors must be communicated to all employees. Engagement of employee discipline must align with company values, promote the right attitude, and be deployed with the authentic commitment and support of leadership. Once acknowledged as an important practice within operations, discipline must be maintained through monitoring and corrective actions when observing deviations from expected norms.

Sustainable Discipline

Operational discipline starts by understanding and articulating what employees are expected to do. Written procedures facilitate consistent and desirable employee behaviors, while institutionalizing best practices and sharing value-added experiences. As indicated, standard work routines (SWRs) provide a degree of clarity and detail behind expected behaviors. People need to be held accountable when not following procedures, especially high-priority, mature procedures critical for the safe and reliable execution of processes and assembly of products. A lack of accountability will eventually lead to deterioration of the required discipline necessary to maintain consistent and reliable outcomes. SWRs help to establish a degree of accountability. Upon monitoring employee performance, a failure to demonstrate discipline for finishing their routines completely or properly needs to be addressed immediately. When chronic underperformance is evident, actions to address the situation should occur in a series of increasingly server steps intended to deter unacceptable behavior.

As a business, if you are experiencing high human error rates, safety incidents, quality and reliability issues, it may be time to consider upgrading or reinforcing discipline practices. In an article titled "Operational Discipline: Does Your Organization Do the Job Right Every Time?" written by Brian D. Rains, Global Process Safety Management Leader, the author describes a process called "progressive discipline", used by many facilities to reflect negative consequences for failure to comply with documented standards. Typical steps in a progressive discipline system are (1) verbal warning, (2) written warning, (3) probation, (4) suspension with or without pay, and (5) termination. He goes on to say *"The discipline system must apply to all employees, from shop floor to shift technicians to supervisors, managers and superintendents. It is most effective when line management is responsible for the communication of the system to all employees as well as its day-to-day implementation"*. [1]

As indicated, the approach described should apply to all employees within an organizational hierarchy in order to establish process integrity and ensure accountability. Once discipline is established, periodic verification that standards are being maintained is an important part of safeguarding the continued existence and effectiveness of organizational discipline. Consequences for failing to follow documented procedures should be clearly communicated and reinforced.

Discipline, complemented by awareness of a long-term corporate strategy, along with the knowledge and skills to realize it, establishes a mindset, the drive, and commitment required to make it happen while accountability ensures operational activities are getting done on time and correctly. It's this type of atmosphere and culture that allows companies to seamlessly move into the future while maintaining competitiveness. Operational discipline is built into manufacturing operations through the application of check sheets, supervision signoffs, and independent verification of critical tasks commonly executed through SWRs and LPAs. Robust operational discipline requires some of the following elements and attributes:

- Management commitment, support, and leading by example.
- A common vision, mission, and organizational alignment.
- SWRs and process audits.
- A desire to achieve operational excellence.
- A high degree of employee morale.
- Transparent leadership practices.
- A do-it-right attitude; no shortcuts, no workarounds.
- An organized and standardized workplace.
- A teamwork atmosphere.

Improving Operational Discipline

In most organizations, operational discipline exists to some degree and is generally a reflection of company size, product complexity, business risk, and management understanding of its value. For example, there is a much greater business risk when making a product for the aerospace industry than for the paper industry. Regardless of industry, there are typical benefits for most companies interested in improving their operational discipline. Success is likely when the willingness and effort to achieve it exists.

A structured approach to improving a sustainable operating discipline may include creating awareness and understanding of its benefits and the drawbacks of not being disciplined. Articulating the risks of not exercising proper discipline should be considered. For example, disciplined practices not exercised by airlines or the nuclear power industry can be disastrous. Where the risks of failure can significantly impact the company, its customers, users, or surrounding environment, it's important to clearly define employee expectations by documenting specific work routines, their priority, and frequency of completion. The consequences of not performing to company standards should also be measured and communicated. When consistent output performance and system integrity are significantly influenced by the degree to which discipline is exercised by an organization, it will be necessary to continuously review employee performance in light of conformance to operating norms.

Expectations for a particular behavior must be clearly communicated to all employees through company policies and procedures in addition to their daily SWRs. Where necessary, timelines should be monitored and realization of deadlines confirmed. Managers have the responsibility to display the desired discipline and solicit feedback from employees to assess their compliance to work standards. They must use this information to evaluate each employee's ability to effectively manage operational activities within defined working parameters.

How to Maintain Operational Discipline

Good process requires good discipline. The more complex a process, the more discipline is required to maintain control and drive improvements. Discipline stems from the standards and methods an organization puts in place to drive business operations. Discipline also requires that employees follow standard as defined. Here are some tips for ensuring operational discipline is maintained with time:

- Clearly define and communicate acceptable work behaviors to all employees.
- Measure, monitor, and control manufacturing practices to reinforce desired behaviors.
- Discipline practices should be tailored and reinforced to the degree needed to ensure compliance to operating standards.
- When instilling or strengthening shop floor discipline, start with basic standards such as workplace cleanliness (5S) or work instruction compliance.
- Leadership and management should articulate and exhibit the disciplines expected of their employees.
- Solicit feedback from employees to confirm their understanding of expectations for exercising discipline in the workplace. Take action to correct any misunderstandings and reinforce desired behaviors.
- Emphasize the practice of operational discipline through the periodic verification of standard work instructions and SWRs.
- Engage in regular performance reviews of key metrics to confirm operational discipline remains sufficient to deliver expected system outcomes.
- Conduct LPAs as an integral part of your operating practices.
- Recognize that some people will never possess the needed discipline to work in a shop floor environment.

In summary, discipline is a mindset which stems from a commitment of management and employees to execute their roles and responsibilities as key contributors to organizational success. Management must ensure the right systems are in place to promote behaviors expected to achieve ideal results while employees must exhibit the discipline required to follow and improve standards, procedures, and work routines designed to help realize and continuously enhance output performance.

Knowing the right thing to do takes knowledge, experience, and an inquisitive approach to understanding and decision-making. Being willing to do right things right takes personal integrity and a willingness to communicate when expectations can't be met due to unexpected circumstances. It also requires coaching and mentoring others to ensure things are done according to company policy, procedures, and standards.

Operational excellence requires operational discipline which relies on the integrity of employees to follow through with their commitments. Integrity is doing what is expected of you and being committed to doing it the right way, whether or not someone is looking. Integrity builds trust, reliability, and confidence as others rely on you to satisfy obligations. Being able to predict how people will behave leads to more accurate planning, reduces waste, aligns activities, and drives productivity. Integrity is an essential component of operational discipline.

SIDEBAR: DISCIPLINE MATTERS

Management by email does not work well. Case in point, I once worked for a large global company that had difficulty changing procedures, not because we were large but because the process was long and "complicated". To work around this problem, management would effect change through emails. Plants would periodically receive an email from the central office stating a change in procedure. The emails could be written by anyone from a project manager to a senior leader. The casual email might "eliminate" a documented requirement, request a change in procedure, or add a new requirement to shop floor activities. The plants were frustrated because they were being held to an industry standard of following approved procedures in order to demonstrate compliance while the central functions expected them to follow the latest email versions with little regard for the formal change management process.

Sadly, instead of improving the change control system, the central team chose to circumvent it by issuing email changes making the process difficult, if not impossible to control. In the middle were the plants struggling to follow their document control system while trying to satisfy senior management's "need for speed". In the end, management believed it was easier and faster to drive their agenda via email than to use the change management system, which they were required to follow. Clearly, this demonstrated a failure of leadership, exhibiting the wrong behavior and mindset for a lean thinking organization.

Key Points

- Operational discipline is about doing the right thing, the right way, every time.
- Operational discipline is an enabler of manufacturing excellence.
- Discipline starts with knowing who needs to do what, when, where, how, and how often.
- Discipline is the practice of people following rules or standards of behavior. It's the attribute of being able to behave and work in a controlled way.
- Management is expected to exercise and instill discipline by coaching and mentoring employees.
- Rules require discipline and accountability to be effective.
- Operational discipline reflects company culture and more specifically employee behaviors.
- Operational discipline becomes more valuable and essential with an increasing number of workers and operational complexity.
- Operational discipline is established through policies and procedures, exhibited by management and exercised through roles, responsibilities, and work routines.
- Military precision requires operational focus and gets desired results when supported by a high-integrity system.
- Greater operational discipline is achieved through awareness and proper execution of documented procedures.

- Improving operational discipline reduces process variation which leads to more consistent and predictable output.
- Operational discipline becomes increasingly more important as the degree of business risk increases.
- Discipline is about employees doing the job right, by following required procedures, and not taking shortcuts to do so.
- Discipline involves a commitment to complete tasks completely and correctly, on time and every time.
- Methods, tools, and techniques such as check sheets, signoffs, and independent verification can help maintain the required operational discipline to ensure process integrity and product quality.
- Organizational discipline is about following operational standards, procedures, and work instructions and updating standards when a better way (safer, cheaper, faster, easier) is found to perform work.

Source

[1] Rains, B. D. (2010, June). Operational discipline: Does your organization do the job right every time? In *Global process safety management leader*, DuPont Sustainable Solutions.

1.4

Employee Accountability

The first and best victory is to conquer self. To be conquered by self is, of all things, the most shameful and vile.

~ Plato

Objective: Create an environment where employees know what to do and have the needed resources to do it. Discipline is the result of holding people accountable for their commitments.

Overview

Accountability is not a term that's easily defined. It's about holding someone responsible for something they are expected to do or have done. Being "responsible", according to Dictionary.com, is *"having an obligation to do something, or having*

DOI: 10.4324/b23307-5

control over or care for someone, as part of one's job or role". Although similar to accountability, the key difference between responsibility and accountability is that responsibility can be shared with others while accountability cannot. Accountable people must be answerable for their own actions. An individual who is accountable is also responsible.

Accountability is about taking ownership. Ownership is the act of possessing something or taking responsibility for an action, activity, task, or project. Ownership can be about gathering thoughts and ideas from others and using this knowledge to realize certain goals or objectives. In society, it can involve working with people to achieve specific outcomes or results, regardless of who did the work. Ownership can be viewed in a broader sense of supporting those working cooperatively together. This perspective can send a message that you are a team player and willing to participate as a team member for the greater good of an organization.

Sometimes, taking ownership may require doing things outside your scope of responsibility to achieve a desired outcome. This may include accepting responsibility for completing a task or assignment that others may not have the time, resources, or interest in doing. This can be a significant boost to the team or organization, especially if the task is difficult or unpleasant. Taking ownership, especially of activities outside your normal job responsibilities, can enhance your problem-solving skills and expose the willing individual to new challenges, opportunities, and people.

Ownership can be facilitated through a shared vision which allows people to consciously engage on issues that align with the organizational mission. This connection can be enhanced when individuals are encouraged to share their ideas, knowledge, and insights during problem-solving and decision-making. Clarifying the "why" behind certain tasks will allow people to see their value contribution to a group or enterprise, strengthening their commitment of task ownership. Providing employees with direction and necessary authority, while allowing them to decide how to achieve the desired outcome, will promote trust, inspire confidence, and further commitment. This can also serve as an opportunity to coach and mentor employees in the disciplines of problem-solving by asking probing questions that shape their critical thinking process, develop their skill set, and enhance their leadership capabilities.

It's also important to hold employees accountable since employer trust and employee autonomy go hand in hand. Holding employees accountable and expecting their best work helps motivate them to achieve better results. However, accountability requires that expectations are clearly defined, documented, and understood. It also necessitates that an employee can do what they are being asked to do. Exercising accountability is often considered a desirable leadership trait that is often noticed, appreciated, and rewarded in different ways.

Corporate Accountability

Corporate accountability involves being answerable to all stakeholders for an organization's actions and results. Employee accountability, in a corporate setting, relates to an employee's role, responsibilities, and work routines which can be evaluated through an individual's behavior, performance, and achievements. It involves assigning individual ownership to tasks that help maintain a management system and ensure individuals preserve system integrity by delivering consistent and predictable performance. Holding employees accountable means ensuring they are completing all their work on time and correctly while acting to highlight, understand, and address process abnormalities.

Workplace accountability is linked to many different factors, including workplace attendance, completing assigned work tasks or routines, exhibiting certain behaviors, and working together in teams. Thus, accountability is not only an individual responsibility, it can be a shared responsibility that exists between people working together toward a common goal or objective.

Employee Accountability

By our very nature, human beings are flawed; they are not always reliable or capable of delivering on their commitments. People often make mistakes and, as a result, fall short of expectations. One expectation of managers is to hold their employees accountable while taking time to work with them to deliver successful outcomes. The primary objective of employee accountability is for competent individuals to fulfill their work obligations and further the goals of the organization.

Employees are hired to satisfy certain defined or expected work activities for which they should be qualified or capable to complete. High-performing organizations clearly define these activities prior to hiring an individual or ensure the right candidate is hired or the individual selected can be properly trained to do the job. If required tasks or operational duties are not completed, or not completed as expected, the employee may suffer repercussions because of their actions or inactions.

Holding employees accountable is an intentional act that takes time to institutionalize and an unwavering effort to maintain. Accountability to standards, procedures, and results must become an integral part of an organization's DNA. Properly executed, accountability can lead to a culture of empowerment and trust. Poorly executed, it can create an environment of suspicion, micro-management, and distrust. A lack of accountability can eventually lead to chaos, as employees decide what to do and how to do it, within their scope of responsibility.

How management and employees behave in their work environment will significantly impact organizational culture. Accountability must be executed with respect for the individual, displayed through employee trust, and backed by the support of management. This also relies on hiring and developing the right people, setting clear strategic goals and objectives, and aligning employee roles and responsibilities with those expectations.

Maintaining a culture of accountability also requires a continuous effort of assessment and feedback with corresponding rewards and consequences. Consequences may run the gamut of re-training, to disciplinary action, and to termination. Rewards can range from positive verbal reinforcement to financial compensation and promotional recognition.

SIDEBAR: WAYS TO ENHANCE EMPLOYEE ACCOUNTABILITY

- *SMART goals* – establishing SMART goals can be an effective way to enhance employee accountability since SMART stands for specific, measurable, achievable, results-orientated, and time-bound. These characteristics clarify and articulate reasonable expectations, leaving little wiggle room for irresponsibility.
- *Monitor time and attendance* – knowing who was where and when during a shift adds a layer of connection to work and related events. People tend to be more accountable when their whereabouts and work performance are tracked.
- *Team motivation* – when teams share compensation risk and reward, peer pressure can enhance employee accountability. Motivation by one's peers can be a powerful tool for achieving expectations.
- *Prioritize activities and actions* – recognizing and continuously driving priorities can be used to maintain accountability. Prioritizing work responsibilities and actions provide focus and a sense of urgency that often can't be ignored.
- *Performance reviews* – people are interested in doing a good job, especially when they know their work activities are important and are being reviewed with the support of management. By default, monitoring work progress promotes increased productivity and accountability.
- *Recognition and reward* – people respond to positive feedback and tend to become more motivated when they feel their work is meaningful and appreciated. Recognizing people for their activities acknowledges their work and reinforces their accountability.

Peer Accountability

The individuals who people work and interact with daily often have a better understanding of each other than the leaders who manage them. Time spent working with and among others can reveal a lot about individuals and their work habits. This level of closeness and insight can create awareness of how people work individually and in a team environment. Many self-organized and directed teams can create an atmosphere of peer pressure that regulates performance and accountability within the workplace. In tight-knit organizations, accountability is frequently driven by peer pressure more than management, especially when the work environment requires people to work closely together. This can play a significant role in the attitudes and behaviors of employees. Therefore, building strong working relationships within self-motivated teams is paramount. It's important to note that teams can play a significant role in influencing organizational discipline and employee accountability.

Shop Floor Accountability

Nothing happens on the shop floor without knowledgeable and experienced people committed to making it happen. As a result, the shop floor must be staffed by competent individuals willing to follow instructions, able to manage people, properly handle materials, diligently maintain equipment, and actively control the work environment. Some of the more fundamental activities include employees implementing methods, procedures, and work instructions, as well as solving problems and improving processes. Managers are expected to coach, mentor, and guide people through their daily work routines to maintain and streamline operations. Material is employed to make parts and assemble products. Machines, tools, and gauges (e.g. equipment) are available to facilitate production. Environmental factors are controlled to maintain a comfortable and stable working environment while information systems are used to manage people, execute processes, supply materials, maintain equipment, and assess inventories. These are only a few of the many activities required for effective shop floor management. If planned and executed well, many benefits can be realized, including the production and delivery of cost-competitive products, continued operational stability, and productivity improvements. A reduction in work errors, process deviations, and product defects may also be observed, resulting in an upward trend in manufacturing revenue.

Unfortunately, none of this will happen unless people are held accountable for the processes, procedures, and systems put in place to realize these benefits. In other words, you can have the best systems in place but if you don't have the operational discipline to follow them or the accountability for ensuring they were implemented correctly and completely, you are unlikely to achieve excellence in manufacturing operations.

Accountability is best achieved when transparent processes are in place. This allows for rapid and continuous confirmation that an operating system is being executed as intended and continues to deliver consistent and predictable results. Shop floor accountability can be exercised in many ways. Tips for exercising shop floor accountability include the following:

- Periodically verify standard work is being followed.
- Confirm process outputs are meeting targeted results.
- Review data for expected performance and trends.
- Conduct audits to validate system integrity.
- Walk the shop floor looking for abnormalities.
- Confirm incoming materials comply to specifications.
- Review customer returns and feedback for improvement opportunities.

Setting Employee Expectations

Setting expectations at the start of employment is a good way to introduce individuals to organizational culture. Spending time cross-training in different departments or several weeks with a mentor, prior to starting in a new position, can provide insight about how the company works and operates. This experience sets the expectations (framework) for employee behavior and accountability. Let's consider several ways to establish discipline and accountability in the workplace.

Employee Feedback

An important part of developing and reinforcing employee accountability is reflected in the frequency and timeliness of actionable feedback provided to employees. People need to know if they are meeting organizational expectations, and if not, what adjustments or changes they can make to satisfy those expectations. Feedback, done right, will enhance a sense of belonging and value for work performed. Having a meaningful employee feedback system, that makes it easy and efficient to provide feedback, will enhance the work environment by showing employees that the company values their contribution and is interested in their personal growth and professional development.

Employee Empowerment

Freedom to work within clearly established boundaries creates trust, motivation, and employee engagement to act creatively, yet mindfully, when fulfilling their responsibilities. Empowerment stems from the trust management bestows on their employees to do the work expected of them while openly encouraging them to ask questions and request help, when needed. Empowerment is "given" by management but must be continuously "earned" by employees. Giving employees the ability to take ownership of their work and make decisions, within their scope of responsibility, is a key step in being able to hold employees accountable for their actions.

Rewards and Consequences

Every employee should be clear as to the rewards and consequences of their actions or inactions. Accountability requires that employees understand company expectations and the consequences of not living up to those expectations. On the other hand, good performance should be rewarded to reinforce the desired behaviors expected of individuals. Management sets the tone in responding to employee actions and should be aware of how their response influences company culture. Fairness in exercising employees' accountability reflects the consistency in which rewards and consequences are enacted across the organizational hierarchy.

Treat People with Respect

Accountability is best established and maintained when issues of concern are treated with honey versus vinegar. Dealing with concerns head-on will show employees the company is serious about holding people accountable for their behaviors and actions. However, doing so respectfully allows for an open and honest interaction and facilitates the retention and development of valued employees. Holding a face-to-face conversation, in a neutral environment (e.g. coffee shop), can minimize defensiveness and promote a receptive and engaging exchange. The objective is to have the employee leave the discussion with a renewed understanding of expectations, a sense of being valued, and a strong commitment to their work. The consequences of not meeting expectations should be discussed openly, honestly, and fairly with clear actions to achieve the desired behaviors.

Communication

Good, on-going communication is key to establishing and maintaining an accountable organization. If people are unaware of the expectations and activities impacting employee roles and responsibilities, it's difficult to hold them accountable. A good communication strategy will help keep needed information flowing to the right people in a timely manner. It will also allow employees to make appropriate adjustments, in their daily work routines, to effectively control, improve, and optimize activities within their current scope of control. Accountability is more than just doing the work; it involves situational awareness and being able to adjust one's work activities and habits to continuously improve and optimize performance.

Strategy Alignment

Accountability is most effective when employee expectations are aligned with the organizational strategy. When common goals and objectives are effectively deployed throughout the enterprise, realization is closer at hand. Operational efficiency also improves the more aligned the employees are with the strategic direction. Maintaining accountability to common goals and objectives demonstrates alignment with a shared purpose. Success in today's world is about remaining ahead of the competition. An organization is expected to advance more quickly toward its goals when everyone is moving in the same direction with clear intent and steadfast resolve.

There are many people and tasks demanding our attention throughout a typical workday, sometimes making it difficult to determine our latest or most pressing priorities. The benefit of having a common, articulated, and well-communicated approach to work is in the simplicity it provides when determining what priorities to focus on next. To keep it in perspective, strategic alignment sets the direction while accountability keeps the team focused, motivated, and moving along the path to operational excellence.

CASE STUDY: ARE YOU A MANAGEMENT TOOL OR EMPOWERED?

I once had a manager who took little interest in what I had to say or thought about my job. I learned over time that I was simply a tool to carry out her bidding and that of her manager. I remember becoming extremely frustrated in meetings when my concerns and suggestions were tolerated but essentially ignored. Although input was often requested, I quickly learned to "shut my mouth" because the request for feedback was simply a formality. Any attempt to prompt a "spirited discussion" on a current topic resulted in an attack on my perceived attitude of not being a "team" player. This meaningless request for feedback gave my manager the license to say she obtained "participation and buy-in" from her team on the decision.

My frustration would span a week of meetings in Germany when I was required to travel to Europe, once a quarter, for global team meetings. As a coping mechanism for not being heard, I started writing. I would use my knowledge and experience to articulate, how I would solve a problem, deploy a new initiative, or address an existing concern. This persistent effort to be "heard", via my writing, eventually turned into my first book *Sustaining a Culture of Process Control and Continuous Improvement: The Roadmap for Efficiency and Operational Excellence*. In essence, empowerment is a gift, not a right, when working for a company or manager. If you are given the gift of empowerment, embrace it since these gifts don't come along very often.

Accountability vs. Micro-Management

There is a fine balance between accountability and micro-management. The complexity of that balance becomes more apparent when individual personalities are involved. At one extreme, considerable attention must be given to certain individuals who require constant direction and engagement to get things done. At the other extreme, a simple request may be all that's needed for knowledgeable, experienced, and motivated individuals to deliver on a request or assignment only stopping occasionally to ask questions or obtain clarification on the expected outcome.

It's important that those holding people accountable are aware of how their employees work and what motivates them to deliver on their commitments. Time and attention spent on following up with team members and direct reports must be commensurate with the attitudes, behaviors, and engagement demonstrated by each employee. The perception of micro-management can be damaging when interfacing with highly energized and motivated individuals.

When a high level of engagement is required with an individual, be polite, persistent, and patient since it may take time for an individual to recognize and understand organizational expectations. Over time, mindful employees will eventually learn what's acceptable behavior for their roles and responsibilities. If, after a given period, certain individuals continue to display "high-maintenance" behavior, moving them to another position or out-the-door may be an appropriate response.

Attitude, Behavior, and Accountability

Attitude is a way of thinking or feeling about someone or something and is typically reflected in one's behavior toward others and their own work habits. An employee's attitude plays a significant part in personal and organizational accountability. It's important to identify, hire, and retain highly-motivated employees. In most instances, this is relatively easy since motivated employees typically know what drives them and can articulate what they need to remain engaged. In those instances, where a manager can no longer offer an employee what they need, it's generally in the best interest of the company to find new opportunities for these individuals, within the organization, since acquiring and retaining talented and valued employees is difficult at best.

One's attitude in holding an employee accountable must be that of respect for the individual. People respond more positively to those they trust and respect. An open and honest discussion of a situation or assignment and clarification of an employee's role or impact on its outcome may be all that's needed to reinforce the importance of meeting due dates or completing daily work routines properly and completely. Always retain an air of civility and respect when interacting with colleagues. Respect begets respect.

Commitment and Accountability

An individual's character can be a good indicator of their ability to make and keep a commitment. A personal commitment is one in which we agree to do something that is not dependent on the help of others. A commitment is a promise to do something: freedom to act without conditions. Accountability starts with a commitment. It's the basis for holding someone accountable. A verbal commitment is stronger than a silent one while a public commitment is even more difficult to break. If you want to acknowledge and reinforce a person's likelihood of keeping a commitment, have them verbalize it. Even better, have them verbalize it publicly.

In certain situations, commitments may need to be monitored since blind faith in an individual or team's ability to effectively follow through with the commitment may be risky. One of the best ways to manage the risk of falling short on a commitment is to revisit it often and remove any issues or obstacles that may threaten its realization. Continuously discussing potential issues and addressing obstacles can help keep a commitment on track for adequate and timely completion. In short, trust in people but reinforce one's commitment periodically!

In summary, employee accountability is evident through the timely execution and completion of employee work activities and is considered the glue that keeps the manufacturing system stable and efficient. System performance must be constantly assessed through periodic verification of workforce conformance to standards and procedures while being reinforced through corrective and preventive actions. In essence, operational performance is rooted in organizational structure, operation discipline, and employee accountability.

Key Points

- Accountability involves assigning responsibility and verifying people deliver on their commitments.
- People must have responsibility for something in order for them to be held accountable.
- If you want to hold people accountable, document your expectations for all to see and understand.
- Workplace accountability is essential for the success of shop floor management.
- People who take ownership can differentiate themselves from the crowd and become the "go-to" person in an organization.
- Employees who work together toward common goals facilitate a more productive and efficient work environment.
- Accountability strengthens when work performed can be linked to the individual who is responsible for it.

- If you are an individual contributor in a company, demonstrated accountability enhances your perceived value as an employee.
- A manager who demonstrates accountability will likely solicit the same behavior from their direct reports.
- A major part of reducing human error is to make everyone accountable for their actions.
- Hold people to their commitments! Trust, but verify! Be polite, pleasant, and persistent!
- Employee accountability includes all employees, from the maintenance team to the CEO of the company. Every employee has a role to play in the organization which should be clearly defined and aligned with the enterprise strategy.
- It's the responsibility of employees to complete the tasks they are assigned and to perform the duties for which they were hired.
- Personal accountability involves the willingness to stand behind your decisions, actions, and behaviors. When you're personally accountable, you stop assigning blame and making excuses; you take personal responsibility.
- Personal accountability means owing your assignments and following up on commitments.

SIDEBAR: PERSONAL ACCOUNTABILITY

Accountability starts with oneself. If you want to enhance your personal accountability, consider some of the following points:

- *Prepare a mission statement* – identify what is important to you (e.g. what are your values and aspirations?). Write them down and work on living them "a little" each day. Work to assimilate your values into your daily work habits and activities.
- *Set personal goals* – write down your goals, large and small. Make them tangible and measurable. Create a plan to achieve them. Execute your plan at a comfortable pace. Review them periodically.
- *Prepare checklists* – a list can be created for home, work, or other objectives. A checklist is a helpful reminder of what's important or simply what needs to get done. Allow them to guide you, not control you. Prioritize the list. Determine what needs to get done and when.
- *Hold yourself accountable* – manage your time wisely. Reflect on what you accomplished at the end of each day. Consider what needs to be done tomorrow and next week.
- *Manage your time* – allocate a time or a "time slot" to get certain things done. Take stock of your accomplishments, big and small. Limit email and individual distractions. Set aside quiet time or a quiet place to briefly work each day. Close your office door, put the phone on mute, place a "do not disturb" sign on the desk, or work from home one or two days a week. Look for ways to periodically de-stress.

- *Reward yourself* – periodically do something fun or different to reward your accomplishments.
- *Continuously learn* – everyone has strengths and weaknesses. Exploit your strengths and develop your weaknesses in areas of interest. Knowledge and experience enhance your personal and professional value and elevate self-esteem.
- *Mind your time* – time is finite and increases in value the older you get. Balance work with pleasure and other personal interests. All work and no play will impact personal performance. Focus on excellence, not perfection.
- *Seek feedback* – embrace personal performance feedback. Be open to constructive criticism. Obtain multiple perspectives for more valuable insight into your strengths and weaknesses. Keep an open mind and use what you learn as an opportunity to enhance your personal growth.
- *Self-assessment* – periodically review your growth and development plan. Have any priorities changed? Are any updates required? Make necessary adjustments to keep on the path to personal goal achievement.

SHOP FLOOR FUNDAMENTALS

Success is neither magical nor mysterious. Success is the natural consequence of consistently applying basic fundamentals.

~ Jim Rohn

Overview

Once a solid operating structure is defined and deployed, preparing for effective shop floor management requires the implementation of several key processes and procedures to establish a baseline and provide the necessary information for continuous process control and improvement. This baseline includes management system documentation designed to direct work activities, monitor work performance, and manage significant deviations while keeping stakeholders "in the loop" on operational performance. These activities are all executed and coordinated by people with the knowledge and experience to deliver quality product, in the right quantity, and on time to customers.

In the following chapters, we will discuss the methods, tools, and techniques used to help achieve, maintain, and improve manufacturing performance. Standardization is one of the essential practices to control product and process variation and drive consistent output performance (Chapter 2.1). Data and Information (Chapter 2.2) serves operations by providing management system personnel with the facts and figures they need to schedule production, move material, facilitate changeovers, and address abnormalities. These are only a few of the many other things they are tasked to do. In this chapter, we will discuss data gathering, analysis, and reporting of key process indicators (KPIs) used to measure, monitor, and trigger actions needed to maintain process stability and acceptable output performance.

Chapter 2.3 looks at visualizing the data and information gathered and analyzed so it can be easily understood and acted upon, if necessary. This chapter includes a discussion on visual controls used to facilitate the display of KPIs for easy viewing

DOI: 10.4324/b23307-6

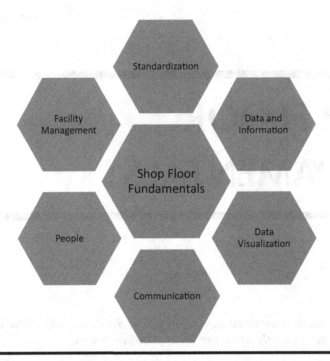

Figure 2.0.1   Fundamentals of Shop Floor Management.

and quick response to process abnormalities threatening productivity. Chapter 2.4 is focused on effective shop floor communication needed to ensure timely data and essential information flow to the people who need it for process execution, monitoring, and control. To make all this happen, you need competent, capable, and committed people to effectively delpoy the management system which is the topic of Chapter 2.5. The final chapter of this section ends with Facilities Management (Chapter 2.6), which is required to create the right environment for the people, equipment, methods, and materials needed to run the shop floor and make the products customers want at the quality and cost they expect. Let's begin our exploration of shop floor fundamentals with a discussion of standardization and its key role in process control and continuous improvement. Figure 2.0.1 captures the fundamental elements of this section.

2.1

Standardization

If you think of standardization as the best that you know today, but which is to be improved tomorrow – you get somewhere.

~Henry Ford

Objective: Implement operating standards to minimize product and process variation without compromising manufacturing flexibility or agility.

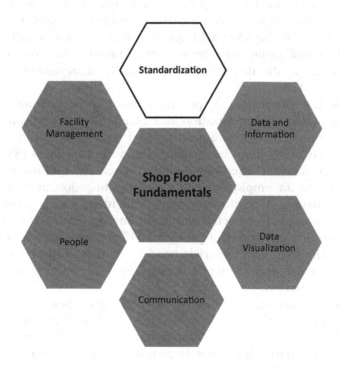

DOI: 10.4324/b23307-7

Introduction

Standardization is the act of conforming to a standard, a standard being a level of quality or attainment. In manufacturing, standardization is often used to define a minimum level of performance in alignment with stated requirements. Standardization is a way to control process variation so that consistent and predictable products and services can be manufactured and delivered as planned. Specifically for manufacturing, standardization includes conformance to processes, procedures, and operating parameters to achieve stable output performance over time.

Standardization is not strictly about following the procedure. It's about executing procedures or work instructions based on the best-known way to complete a process at the time. If a better way is identified and proven effective, the standard should be formally changed to conform to the new or improved way of doing work. In essence, standardization is as much about following procedures as it is about exploring new opportunities to improve upon existing work performance.

Standards also act as the basis for effective process monitoring and control. They allow for a comparison between actual and expected performance. If current performance meets expectations, all is well. However, if a significant deviation from a standard is detected during process monitoring, this should trigger an investigation as to why and action to correct the situation, if justified. Without a standard or target in place, control would not be possible, leading to process deterioration and, eventually, chaos, if nothing is done.

Standards also serve as a reference point for continuous and sustainable improvement. It's impossible to measure improvement if there is no baseline or reference for comparison. Standards serve as a process performance baseline. Improvements can only be sustained if the level achieved becomes the new standard for process monitoring and control. Continuous improvement involves the constant elevation of current performance to the next level of sustainable achievement. This concept is depicted in Figure 2.1.1.

Standards are the foundation for process control and continuous improvement, upon which manufacturing optimization and excellence are realized. Without standards, control, improvement, optimization, and excellence cannot be achieved. Standards often start with shop floor cleanliness and orderliness (5S), since a dirty and unorganized workplace often breeds indifference and inefficiency in work performance. Promoting an employee mindset of optimization and excellence starts with leadership creating a comfortable environment in which people are willing, eager, and proud to work. If expected and maintained, a clean environment becomes contagious. An orderly environment promotes efficiency as people are not wasting time looking for the tools and materials need to do the job they were hired to do. Standardization can also play a key role in ensuring a safe and hassle-free work environment for all employees.

In a robust standardized work environment, supervisors continuously verify employees are following their work procedures. One attribute of an operationally excellent organization is evident when employees notify their managers whenever they can't follow work standards. Immediately communicating this issue obligates leadership to evaluate the situation and act to eliminate the obstacle or create a workaround

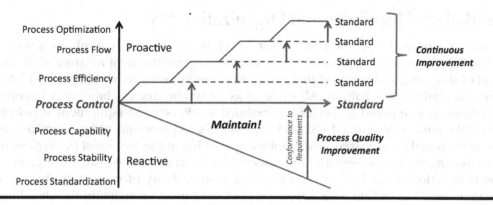

Figure 2.1.1 Shop floor management control and improvement.

until standard work can resume. This is an example of a highly functioning organization and the essence of process control in a way that prevents defect generation by stopping work immediately before it has a chance to be performed incorrectly and potentially create defects.

A continuous effort to confirm standard work must be part of a supervisor's shop floor work routines. This will ensure that deviations from standards are detected early and addressed immediately. Many deviations can be corrected at the time of detection since employees may simply need some additional coaching on the how and importance of following instructions. Sometimes, a containment or workaround is initiated to continue work to meet the schedule. In these cases, the workaround needs to be evaluated for its impact on quality, and a corrective action must be identified to eliminate the workaround as soon as possible so that the intended work rhythm can resume. Consider performing periodic compliance checks during Gemba walks or layered process audits to confirm standard work continues to be followed upon eliminating a workaround.

Processes, procedures, and work instructions are the backbone of any manufacturing operation. These are the activities that make an organization unique and are likely to be your most important assets, next to the people who implement them. This tribal knowledge about production and how it works should be captured, protected, and enhanced through standard operating procedures. It's important that this knowledge does not walk out-the-door whenever an employee decides to leave the company. Retaining vital information should be a top priority and realized through documented work practices.

Standardization should start at the most basic level of operations, ensuring a clean and organized workplace. Standards such as maintaining a defined cycle time for repetitive work may be difficult to achieve, if operators are spending too much time looking for, repairing, or cleaning the items they need to execute their work activities. With this in mind, let's discuss a standard method that many organizations use to establish and maintain a clean and orderly workplace, namely 5S.

Workplace Cleanliness and Organization (5S)

A well-functioning operation starts with a clean, organized workplace. Although "clean" is a relative term, "clean" in a manufacturing environment means a work area void of dirt, dust, and particles that can compromise employee health, product integrity, and equipment reliability. All work areas should be free of clutter: any materials or objects not required to perform scheduled work. Needed equipment should be available, have a designated location, and be in good working order. Standards for the area should be prepared, and employees working in the area must be responsible for ensuring the cleanliness and orderliness standards are met before, during, and at the completion of work. They also have the responsibility of contacting the appropriate individuals or functions whenever they can't meet or maintain the cleanliness standards expected for work. 5S is a popular methodology for helping to establish and maintain a clean and organized workplace. 5S can act as a bellwether for the discipline an organization needs to formulate a lean attitude and excellence mindset. It establishes an initial baseline for process stability and control. 5S stretches beyond the basic concept of housekeeping; it is the foundation for process standardization, which will be discussed in more detail in Chapter 2.

Standard Procedures and Work Instructions

Following a methodology for workplace organization is the first step in achieving manufacturing excellence and, as stated, the foundation for process standardization. The next step typically focuses on standardizing critical manufacturing processes and procedures through documented standard work instructions (SWIs). Standard work captures the best-known way to perform value-creating work on the shop floor. It reduces process variation, aids in process stability, and establishes a baseline for continuous improvement activities that can lead to work optimization. At a minimum, SWIs should define the activities to be performed, the sequence of those activities, the cycle time for completion, and expect results upon completion. Compliance to SWIs can be reinforced through periodic checks and process audits as part of the manufacturing management system. Changes to standard work procedures and instructions should occur using a robust change control process.

Every documented procedure needs to reflect the best knowledge and experience available to ensure its timely execution yields consistent and predictable results. Properly defining how a task should be performed, every time, is critical since following a poorly documented procedure can degrade process performance. Operational excellence requires high-quality procedures executed with integrity and military discipline. Figure 2.1.2 provides a visual example of a standard work instruction.

Standard work instructions (SWIs) describe work that is highly specified as to content, sequence, timing, and outcome. SWIs should address the work assembly sequence, handling of non-conformities, material supply requirements, tooling, and documentation management. At a minimum, SWIs must be prepared for manufacturing assembly operations. The following are guidelines for preparing SWIs:

Standard Work Instruction

Workplace: ABC Index 01

Status	Released	ID	Operator	Industrial Engineering	Line manager	Quality	Plant
Released	18.08.2017		Jackson / Larry	Samual / Reese	Harris / Sally	Hunor / Ralph	Auditor: Randolph / Matt

Last modification: 08.08.2017

Key symbols: Tools · Safety · Ergonomics/Health · Visual inspection · Quality / Special Characteristics · Manual test

2 Op.	3 Op.	N+1 Op.	Symbol	Serial #	Elementary operation	Pic. Nr.
1	1	1		1	Take one PCB with right hand & one housing with left hand. Put PCB into housing.	1
1	1	1		2	Take away finished part with left hand and put assy part on heat stacking machine with right hand. Start machine with right hand.	2
1	1	1		3	Unload machine with right hand and put finished part in waiting position in front of machine.	3
1	1	1		4	Load part to loading position with left hand on assembly machine. Take one component with right hand and put it in loading position inside of	4
1	1	1		5	Pick unit with right hand and walk to glue station	5
1	1	1		6	Unload with left hand and load part to loading position with right hand on glue machine. Start machine.	
1	1	1		7	Walk to next station.	
1	1	1		8	Check solder ball with magnifier; clean the part if there is solder ball. More details see attachment 1 "Visual inspection soldering"	6, 7
1	1	1		9	Walk back to first station.	

Pictures (numbered 1–5 in the right column)

Figure 2.1.2 Example of a standard work instruction.

✔ Include safety and ergonomic elements.
✔ Define operational steps (elements) in detail, including right-hand and left-hand movements.
✔ Include detailed pictures and other visuals.
✔ Descriptions should be clear and concise.
✔ Include customer-specific requirements and symbols.
✔ Photos should be large enough to clearly visualize the points they are demonstrating or reinforcing.
✔ Font size chosen for the SWI should allow everyone with 20/20 vision to comfortably read the document.

To achieve standard work, a process should be repeatable, reproducible, reliable, available, and capable of delivering consistent production quality.

Standard Operation Sheet

In a high-volume manufacturing facility, the "standard operation sheet" or SOS is used to document the number of operators within a production line and their workflow, along with the cycle time expected to complete a full cycle (one unit). A standard operation sheet (SOS) should be available for every production line where standard work is expected to be performed and maintained. This document, along with SWIs, can be referenced for periodic confirmation that standard work continues to be executed as defined. A separate SOS should be available at the production line for every operator configuration since the flow of work and cycle time may change as the number of operators changes. At a minimum, the SOS should display the production line's planned cycle time, production line layout, work sequence, and the position and motion of operators during the standard work cycle with and without parts. See Figure 2.1.3 for an example of a standard operation sheet. The following are guidelines for preparing a SOS:

✔ The SOS should be aligned with the SWI steps for easy reference.
✔ Consider using solid arrows to visualize operator movement *with parts* and dotted line arrows for operator movements *without parts*.
✔ Optional additions to the SOS include line-material feeding systems, buffers and position of material stock, and standard work-in-process.

Workplace Design

Workplace design involves the activity of designing and organizing a workplace optimized for worker performance and safety. It may include attention to aesthetics, space planning, and functionality. In manufacturing, efficiency and ergonomics must also be considered in workplace layout. A good workplace set-up will focus on the workspace, applicable tools, and body position to facilitate material handling with minimal operator bending, twisting, and stretching, to reduce the likelihood of

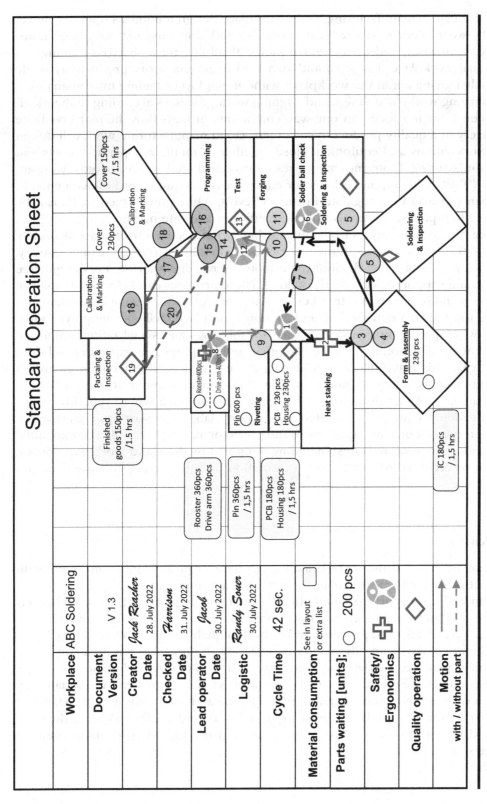

Figure 2.1.3 Example of a standard operation sheet.

personal injury. In manufacturing, good workplace design minimizes operator movements between work steps, reduces inventory and wait time between operations, improves teamwork, and increases workforce flexibility through cross-training. An exceptional workplace design should also help managers verify productivity while seamlessly moving about the workplace without employees feeling micromanaged.

Optimizing workplace design and aligning work processes according to their logical sequence, with a focus on one-way continuous process flow, helps to facilitate the delivery of a quality product to customers. Good manufacturing work cell design keeps workstations and equipment close together to minimize operator movement and manufacturing footprint, minimizes non-value-added space, avoids work-in-process (WIP) build-up, and allows for easy access to equipment for maintenance. A Kanban can also help regulate inventory levels, while gravity feed racks can be employed to replenish materials in a more efficient and safe manner.

Office design also applies to manufacturing facilities since office space is usually a part of the factory layout. Office workspaces should encourage team creativity, provide appropriate work flexibility, promote knowledge sharing, encourage free thinking, and support calculated risk-taking. These practices can help with employee satisfaction and retention. Another key objective of workplace design is to retain and leverage top talent. A comfortable and friendly workplace facilitates employee satisfaction. When preparing a place to gather for meetings and socialization, employee interaction and collaboration should be considered to optimize engagement and team productivity. This may include considering the use of wearable technology, media streaming, wireless charging, and cloud-based personal communication services.

In summary, standardization can be applied to any area where repetitive work exists, including equipment maintenance, quality labs, line start-up, changeovers, and report generation, among others. Standardization practices can also be expanded to a broader scope of activities, including support functions and workplace layout, with the objective of reducing process variation and generating consistent output performance.

Key Points

- Standardization is the act of bringing a process or product into conformity with a set of requirements to ensure consistency and regularity.
- Standard work is primarily intended for operators to perform their job functions correctly, not for a supervisor to confirm that the operator is following it.
- A standardized work process can be used to qualify employees to do a job.
- When deployed correctly, standards will help the organization focus on value-added work, reduce defects, avoid rework, and minimize cycle times.
- The Plan-Do-Check-Act cycle can be an effective way to deploy standards by first defining the standard (plan), then training and communicating the standard (do), followed by auditing the standard (check), and finally reviewing the standard for effectiveness, identifying "gaps" and making adjustments to improve the standard (act).

2.2

Shop Floor Data and Information

Without data you're just another person with an opinion.

~Edward Deming

Objective: Collect reliable data to prepare good information for awareness, problem-solving, and decision-making.

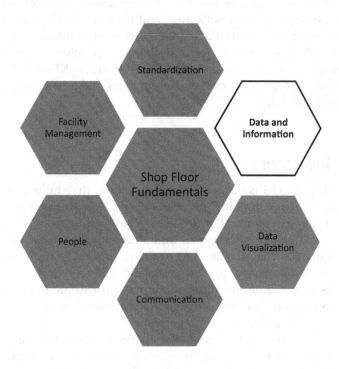

Measurement Data

Data are the lifeblood of any manufacturing operation. Data provide the information required for process control and the insight needed for problem-solving and decision-making. Data can result from measurements used to monitor process performance. Measurements imply a scale and can be used to justify a rationale, make a comparison, or indicate a performance level. We measure to establish a baseline against which process performance and improvements can be assessed. Measurement data can lead to the establishment of metrics which can be used to influence behavior and confirm implemented improvements. From a shop floor perspective, a metric is a standard of measurement that reflects process performance in relation to a target.

There are two types of measures: quantitative and qualitative. A quantitative metric is numeric and describes a measured value on a continuum or scale. Height, weight, and diameter are examples of quantitative metrics, whereas a qualitative metric is non-numeric, such as small, medium, and large, or good versus bad, and are sometimes described as a rank-order measure of value with a limited number of levels. A good metric is SMART, which stands for simple, measurable, attainable, repeatable, and time-based.

Metrics are often used to promote awareness and enhance knowledge of process performance and improvement. In a production line, when monitored closely, and responded to quickly, metrics can help reduce the impact of poor performance on subsequent manufacturing processes that rely on product quality from upstream workstations. Metrics can quickly become overwhelming, if there are too many or there is no clear strategy for data collection and use. Due to the vast amount of data and information available in today's manufacturing environment, it's important to know what critical data are needed for daily process monitoring and control. This is often accomplished by identifying key process indicators (KPIs), which will be discussed next. A table of typical manufacturing metrics and a brief description of each are provided in Figure 2.2.1.

Key Process Indicators

Key process indicators (KPIs) are measurable values that help determine a company's effectiveness in maintaining process control and sustainable improvements. They are a set of measurements that quantify results and can be used to evaluate a company's ability to achieve operational and strategic objectives. They are typically critical (key) parameters that are used to evaluate process compliance and achievement. KPIs can include quality, productivity, and delivery performance indicators. KPIs provide a focus for strategic and operational improvement, create an analytical basis for decision-making, and help focus attention on what matters most. As stated by Peter Drucker, "what gets measured, gets done".

KPIs can be used to track progress when complemented with targets. Influencing performance can occur when leading indicators, such as process input parameters, are manipulated and controlled in a way that affects output results. A good process indicator has some of the following attributes:

Metric	Description
Cycle Time	the total time spent working to make a product and can be calculated by taking the total amount of goods produced divided by production time
Defect	a deviation from specification or target, a non-conformance to expectations
FPY	First Pass Yield is a measure of process throughput or performance and is the percentage of good parts made without any rework or repair
Lead Time	the time between the beginning and end of a production process
Output Quantity or performance	The number of good product yielded from a process
OEE	Overall Equipment Effectiveness is an industry standard of measurement of production efficiency that results from equipment availability, output performance and product quality
Labor efficiency	a measure of laborforce efficiency relative to a standard; actual output versus planned output of a production line
Scrap	a nonconformance to requirements due to manufacturing defects or errors resulting in discarded material
Rework	a repaired part, product or component due to a manufacturing nonconformance to standard
Waiting/ idle times	the time employees wait for work due to issues such as a machine down or missing material. It's calculated as the scheduled production time subtracted by the actual production time
Work in process (WIP)	amount of work in production waiting to be completed
On-Time Delivery	the rate of finished product delivered on time

Figure 2.2.1 Typical manufacturing metrics.

- Provides objective evidence of progress toward achieving a desired result.
- Provides what's needed for better problem-solving and decision-making.
- Can be used to assess performance trends and process change over time.
- Provides visibility into predictable output performance.

KPIs can take on several characteristics. They can be qualitative (e.g. pass/fail) and quantitative (numeric) in nature. They can act as leading and lagging indicators for process monitoring and control. Leading indicators can predict process outcomes while lagging indicators can reveal an outcome as a success or failure. Operational intelligence can be significantly improved when an organization understands how various metrics are used and how different types of measures influence shop floor performance. KPIs can be categorized into three types: inputs, process activities, and outputs. Input and process activity indicators provide insight into output performance. Input indicators include parameters such as amount, type, and quality of material. Process activity indicators are parameters that can only be monitored or controlled at the time of production, such as process time and temperature. Process output indicators determine resulting process performance. These output metrics or results typically include units produced, scrap, and yield. Figure 2.2.2 depicts the relationship of input, process activities, and output indicators.

Metrics, or KPIs, are used to assess how the process is performing. It's a way of measuring success relative to customer expectations and business value. KPIs help link customer requirements to process performance. They relate actual process performance to standards or targets and facilitate fact-based decision-making. In certain circumstances, they can help prioritize improvement activities. KPIs are a mechanism

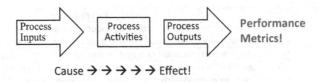

Cause → → → → → Effect!

Figure 2.2.2 Input, process activity, and output indicators.

Element	Term	Description
10	Figure	A value without description; undefined.
	Metric	A system or standard of measurement
 SPEED LIMIT 10 MPH	Target	An objective or goal to be achieved or maintained
Speed	KPI	A measured value used to assess a level of performance or success

Figure 2.2.3 Types of measures.

for assessing the "health" of manufacturing processes. The differences between various terms used when discussing measures are overviewed in Figure 2.2.3.

KPIs help us determine if we are doing the right things right. They measure current performance for comparison to manufacturing objectives and can trigger corrective actions when performance does not achieve desired targets. KPIs often prompt the need for deviation management. To maintain a process at peak performance, KPIs must be monitored daily (if not more frequently) by production teams, reviewed by management during shop floor walks, and discussed during performance reviews. Action should be considered for any KPI below its expected target or whenever unfavorable performance is evident. Monitoring key process indicators is an inherent part of daily shop floor management. It provides focus on the significant few over the trivial many parameters. KPIs are a subset of metrics because they are a select group that serves as a barometer for manufacturing performance.

A KPI should be selected by management to reflect performance requirements and organizational goals. They should provide meaningful insight as to the stability and consistency of the processes they represent. They must be meaningful in that they are rooted in reliable data, easily understood, and can prompt action, when necessary, to address significant deviations. Short-term KPI analysis is used for daily process control and improvement, while historical KPI trend analysis can provide a broader view of organizational performance for achieving strategic initiatives. When posting KPIs for monitoring and control, consider including the following reference information on the charts and graphs displayed:

- Tolerances or target lines.
- Date information was last updated.
- Date next update is expected.
- Name of the individual responsible for preparing, distributing, and maintaining the KPI.

SIDEBAR: SETTING PERFORMANCE EXPECTATIONS: A MANUFACTURING EXCELLENCE MINDSET

Setting performance expectations is not unusual within business operations. Reinforcing these expectations with financial incentives is common practice. This is typically done by defining targets or a value for achieving desired performance metrics. Unfortunately, this approach has a key flaw: once the target or level of performance is reached, there is little incentive for further improvement since the achievement of the target has met management expectations. Across-the-board target setting may also discourage people from focusing on what matters most, since employees may be looking for the "low hanging fruit" to increase the likelihood of goal achievement instead of pursuing the poorest performing lines or processes, which need the most attention.

A better approach, or one that is more aligned with an operational excellence mindset, is to focus on rewarding improvements in performance trends. In essence, employees are not rewarded for achieving a defined goal but for the level of improvement they achieved from a baseline reference value. For example, as the percent improvement from a baseline value increases, the financial reward is commensurate with the degree of improvement achieved. The more significant the improvement, the greater the compensation. In fact, the reward can be based on the value of improvement achieved, allowing the employee or team to share in the level of benefit achieved. This incentivizes employees to continue their improvement efforts beyond a stated goal or target. This approach of trend-based rewards is a lot more realistic, practical, and motivating than trying to reach a predetermined or unrealistic target, both of which may disincentivize employees from continuous improvement, or worse yet, not working on improvements at all.

Data Gathering

We gather data as part of our daily work activities. Data helps define some of the actions that will be taken during the day to maintain and improve operations. These actions may fall within our daily work routines or take us outside our normal work to seek additional information, understanding, and direction. To understand data, we must first make several clarifying points.

A fact is something you observe. Data are from something you measure. A parameter (metric or indicator) is a numerical or measurable factor which helps to define a system or sets the conditions within a process. A process parameter is a performance measure of an activity and a variable used to describe the ***magnitude, dimension, capacity, performance, or characteristic*** of a process, service, or result. It's an **observable** component of data, collected, analyzed, and summarized for decision making and can only be measured in real time.

In manufacturing, data are collected manually and more often automatically using manufacturing execution systems (MES), which serve to facilitate quality and efficient data collection. Unfortunately, data collection can result in wasted time and effort if the data collected is never used. If you are collecting data manually, make sure it will add value in some meaningful way. Consider why data are being collected and how the data are used to support operations. If you can't answer why and how the data are being collected and specifically used, then it is likely a non-value-added activity that should be eliminated.

Data are used to understand process performance and aid in evaluating the integrity of measurement systems. Data can also be used to characterize process stability and capability (Cpk), assess relationships between process inputs and outputs (cause & effect diagrams), and support decision-making based on facts as opposed to guessing or intuition. Data provides the insight management and engineering need to understand current operating conditions and can help direct daily efforts to maintain stability and enable meaningful shop floor improvements. When there is a lot of data to collect and manage on the shop floor, it is recommended that a data collection plan be prepared to ensure data integrity at all times. An example of a data collection plan format and corresponding information has been provided in Figure 2.2.4.

Data Analysis

Data analysis is about taking data, examining it logically or statistically, and preparing meaningful information that can be understood and acted upon, if needed. Statistical data analysis can fall into three categories: descriptive statistics, exploratory analysis, and confirmatory analysis. Descriptive statistics is used to "describe" the characteristics of a sample or data set, such as its mean or standard deviation. Exploratory analysis looks to identify the main characteristics of a data set, while confirmatory analysis attempts to "confirm" or reject an experimental or scientific hypothesis.

It often takes skill to analyze data properly, determine its statistical significance, and present the resulting information objectively. Data can be presented in different ways. It can be summarized in words, visualized in charts and graphs, and verbalized during a talk or presentation. Analyzed data can be used for many things, such as to sway an opinion, draw a conclusion, understand a problem, make a decision, or take action. Depending on the type of data analyzed, the resulting information can be used for descriptive (insight) or predictive purposes, as evident by data trend analysis. In essence, data analysis helps individuals and organizations make sense of data by creating awareness and triggering action. It is important to remember that data becomes meaningless or potentially dangerous if it is collected but not properly understood or used.

Data Collection Plan

Performance Metric	Units	Data Source (Where)	What	Who	When	How	How Much (Sample Size)	Other

Performance Metric – Indicate the type of data to collect

Data Units – Document the metric units (e.g., seconds, grams, inches, degrees fahrenheit, etc.)

Data Source – State where the source of the data are located

What – Describe the data that needs to be collected.

Who – Indicate the person responsible for gathering / reporting the data

When – Indicate when the data needs to be collected.

How – Indicate if any sample techniques or equipment should be used

How Much – Indicate the sample size to be taken.

Other – Include any special instructions (e.g. Frequency of reporting. Who should receive the data, Reporting format & delivery method/location)

Figure 2.2.4   Data collection plan format.

Data Reporting

Data should be available, where and when needed, to facilitate work activities. In modern-day manufacturing, shop floor data can be provided via an electronic dashboard display or more simply written on a board, in real time, close to the line, where the team can use it to plan and exercise their daily work habits. Key metrics should be available for review by anyone at the line, updated on a regular basis, and trigger actions when targets are not achieved, or unexpected trends indicate a problem.

Data reporting should capture the information necessary for process monitoring and control. This may require certain data to be reported more frequently than others. A data communication plan can be prepared to clearly articulate who is responsible for ensuring vital information gets to the right people on time, in a user-friendly and easily understood format. If there is an issue concerning data reporting or the data presented, a clear escalation path should be documented so the responsible individuals can evaluate concerns and act accordingly.

Key Points

- Determine what input, process, and output measures (parameters/metrics/indicators) are required to successfully meet customer and business requirements and manage them!
- KPIs serve as a trigger for identifying process deviations and as the basis for assessing process improvements.
- It is vital that performance metrics are simple, understood, and align with organizational objectives.
- The right metrics will drive efficient and effective shop floor management.
- An individual "owner" should be assigned responsibility for gathering specific data and reporting each KPI.

2.3

Data Visualization

The greatest value of a picture is when it forces us to notice what we never expected to see.

~ John W. Tukey

Objective: Create information from data to control, improve, and optimize the process.

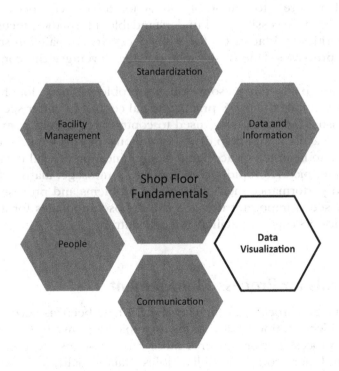

DOI: 10.4324/b23307-9

Overview

The traffic lights and road signs encountered when driving, the dashboard of your car, the numbering system used at a crowded deli counter, and the directional arrows located at intersections, as you hike a complex trail system, are all examples of visual controls. Visual controls can consist of critical data and information that are easily communicated to create awareness and direct decision-making. These controls are everywhere in society and are used extensively in manufacturing to provide key information for assessing process performance and controlling daily operations.

The phrase "Visual Control" describes how key data and information are shared in the workplace to manage operations through process monitoring and control. Real-time data and information need to be displayed or "visualized" so process abnormalities can be recognized, understood, and acted upon. Like the dashboard in a car, the most important operating information should be displayed for quick identification, disposition, and immediate action. Visual controls are critical for effective shop floor management when a fast-paced and dynamic work environment demands rapid dissemination of data and information for problem-solving and decision-making.

When done right, visual controls can be one of the more efficient forms of communication on the shop floor since information of this type can be identified, processed, and responded to quickly. Visual controls should enable employees to easily monitor the process to quickly assess its performance. Visuals can help facilitate the decision to gather more information or take action to correct process abnormalities and deviations. Easily accessible and understandable information focuses people on operational priorities and facilitates the steps necessary to maintain stable, capable, and controlled processes. One of the methods that leverages the concept of visual control is 5S.

In 5S, we clean up a work area so we can see problems since a lot of inventory and a messy factory can hide them. We practice visual control to better see abnormalities "at a glance". Visual controls can be used to control area activities, make decisions, initiate corrective actions, and help drive continuous improvement. Audio and visual aids can be used to heighten or accelerate communication of vital data and information. For example, pictures rapidly communicate a message, alarms draw attention to an issue, and performance metrics highlight problems and process waste. Often, what you don't see is important thus, 5S can expose anomalies for awareness and action. In essence, 5S can also facilitate visual control.

Visual Controls for Process Management

It is easier to manage a process in which key metrics have been visualized. Visualization accelerates the identification of deviations for rapid response to prevent their negative impact on process performance. A more rapid response to problems leads to fewer defects and more consistent deliverables. Data visualization can be facilitated by creating a central area or designated location where metrics can be displayed for periodic monitoring and review by those expected to do so as part of their daily work routines. In addition to being a key tool of process control, data visualization sets the stage for continuous improvement by creating a basis from which progress can be measured and sustainable improvements confirmed.

Visual controls are intended to make process management practices easier and play a part in the continuous improvement of operations. The more visible process performance, the more responsive a manufacturing team can become in exercising process control and improvement. Increasing process transparency allows those responsible for governing the shop floor to be more effective in doing so. Rapid problem-solving facilitates quick resolution of performance issues, minimizing the generation of process waste.

As part of daily operations, actions must be taken to address abnormalities and significant process deviations. The decision to act should be made at the lowest possible level of the organization, where the greatest understanding of the process and corresponding issues typically exist. Effective shop floor management requires real-time, reliable data and information to make quick decisions and identify problems for corrective action. The objective is to avoid defects. For visual controls to be effective in production, time must be spent on the shop floor observing activities, reviewing data, talking with employees, and looking for opportunities to make work simpler, easier, cheaper, and safer for the people doing it.

Through identification and tracking of metrics, a team can follow daily, and even hourly, progress in achieving specific targets, goals, and objectives. The first step is to understand the significant process indicators that influence key operating parameters. Key process indicators (KPIs) are at the heart of visual controls. Selected carefully, KPIs can aid in establishing shop floor priorities that help maintain a well-running manufacturing system.

Although we use the phrase "visual controls", we must consider audio signals as an effective tool for notification of a process deviation or potential concern. We tend to lump audio indicators under the umbrella of visual controls for simplification. However, don't overlook the benefits of using "audio" controls where appropriate and practical. The overall concept of *visual management* is to communicate important information in audio and visual ways. Information is commonly communicated in visual ways, and examples of how this information is sometimes displayed can be found in Figure 2.3.1.

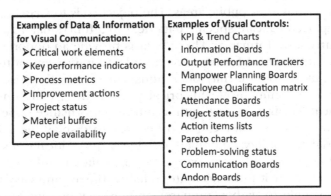

Examples of Data & Information for Visual Communication:	Examples of Visual Controls:
➢Critical work elements	• KPI & Trend Charts
➢Key performance indicators	• Information Boards
➢Process metrics	• Output Performance Trackers
➢Improvement actions	• Manpower Planning Boards
➢Project status	• Employee Qualification matrix
➢Material buffers	• Attendance Boards
➢People availability	• Project status Boards
	• Action items lists
	• Pareto charts
	• Problem-solving status
	• Communication Boards
	• Andon Boards

Figure 2.3.1   Visual communication elements.

As a follow-up to this topic, I would like to leave you with typical examples of metrics that are displayed visually in various areas of a facility to create awareness and trigger action:

- **Safety** – number of near-misses, number of accidents, and consecutive days without an accident.
- **Quality** – number of customer complaints, scrap, rework, yield, and number of customer returns.
- **Productivity** – quantity produced per unit time, cycle time, lead time, machine downtime, and labor efficiency.
- **Cost** – material and labor costs, cost per unit produced, and manufacturing expenses.
- **Delivery** – percent on-time delivery.
- **Inventory** – backlog, turn rate, blocked stock, aging stock, excess and obsolescence.
- **People** – sickness rate, fluctuation (fixed and variable), workforce diversity, training, and attendance rate.

Common Visual Control Displays

There are different ways to display data and information on the shop floor for easy viewing. Manual and digital formats are often used to present information as numeric, graphical, color coded, and as symbols. Some of the more common ways data and information are displayed in manufacturing include information or communication boards (placed on walls or as stand-alone structures near the production lines), manpower planning boards, Andon lights and boards, in addition to matrices to display employee training status and qualifications. Let's look at a few of these in more detail.

Information or Communication Boards

Visual controls need to remain relevant through their accessibility, timely updates, content value, and ease of interpretation. In manufacturing, a common way to achieve these objectives is through the application of information or communication boards placed in common or public areas. The objective is to share general information in a public place or locate specific data next to a functional area or production line to reflect current performance. These boards provide an opportunity to display relevant data and information for process control and improvement to anyone interested in learning more about facility operations, current or otherwise. This information can be used to communicate historical performance, daily output, highlight bottlenecks, reveal KPI trends, and outline future performance expectations based on customer requirements. They can also be used to display the status of human resources, material availability, equipment efficiency, and critical steps for issue escalation and handling scrap. When preparing to post data and information on display boards, it's important for it to be accurate, reliable, timely, and easy to understand. Data should be simple to gather, prepare for presentation, and easy to post and update. If not, it's unlikely to be available or updated when needed. A good information or communication board is:

- Located at the production line or where the information is most needed.
- Provides vital information for managing the shop floor in real-time.
- Clearly highlights deviations from standards or targets.
- Helps establish problem priorities and minimize response time to significant process deviations.
- Allows for organizational transparency in operations by creating awareness.

A typical example of an information or communication board is provided in Figure 2.3.2. An example of a production line board hourly output template is provided in Figure 2.3.3.

Manpower Planning Boards

Manpower planning boards (MPBs) can be an effective visual tool when limited time or resources exist. This tool is often used in agile environments when priorities can quickly change in light of the dynamic flow of information. These boards document known tasks needing attention and serve to visualize these tasks for team discussions and decision-making on available resources to assign next. They can be part of a stand-up meeting where teams meet daily to review what has been accomplished since the previous meeting, discuss issues and obstacles to progress, and make decisions on next steps. MPBs serve to visualize resource availability, facilitate communication, and stimulate rapid decision-making in resource deployment.

Employee Cross-Training or Qualification Matrix

A cross-training or qualification matrix is another form of visual control. It transparently communicates employee qualifications to perform certain job tasks such as product assembly, equipment repair, or supervising changeovers based on successful training and skills demonstration. The left side of the matrix (see Figure 2.3.4) documents employee names, while the top indicates the tasks for which they are qualified to perform.

The matrix provides transparency of employee capabilities and can serve as a skills roadmap for what is needed to qualify for future job rotations and open positions. The matrix provides a reference for employee's medium and long-term development by providing a record of their accomplishments and the necessary steps for additional advancement. Management can use it as a tool to evaluate employee availability based on their qualifications for specific job tasks. There are two cautionary items to consider when posting this type of data: It needs timely updates to reflect frequent changes in employee qualifications to avoid worker discontent with outdated information and the plant must remain compliant to legal restrictions around handling personal data. An example of an employee qualification matrix is provided in Figure 2.3.4.

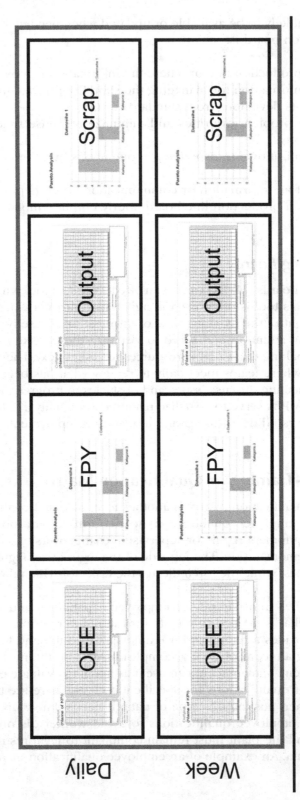

Figure 2.3.2 Information or communication board example.

Time (Hourly)	Model #	Target	Actual	Deviation from Target	Total Quantity Rejected	Problem	Action	Status	Closed by (Name)
6:00 - 7:00									
7:00 - 8:00									
8:00 - 9:00									
9:00 - 10:00									

Figure 2.3.3   A production-line board hourly output template.

Line Leader - Qualification Matrix	Knowledge type / Skill		Knowledge																						
	Training type:			General			Technical									Methodical					Soft Skills				
	Standard procedure:			Shop Floor Fundamentals	Company internal rules	Safety, Environment and Health	Product knowledge	Technical process know-how	Product Quality Standards	Standard Operating Procedures	Work Instructions Standard	ESD Prevention	KPI management	Standard Work Routines	5S BIC	Structured Problem Solving	Kaizen / Poka Yoke	Jidoka	Visual & Deviation Management	Jishuken Workshops	Communication Skills	Stress Management	Assertiveness	Dealing with conflict	Coaching
Div/BU	**Qualification Matrix**		Required Level	B	B	A	B	E																	
Author			Name																						
Check																									
Approval																									
Issue Date																									
Ser. No.	ID No.	Shift																							

Level Description: A = Advanced / B= Basic / E = Expert

Figure 2.3.4 Employee qualification matrix.

Andon Boards and Lights

Andon is the term for *paper lanterns* in Japanese. The application of the Andon concept of lights and signal boards empowers workers to stop production and call for assistance when a process abnormality or defect is found. An Andon board is used in manufacturing to alert management of a problem at a specific workstation. Upon issue identification, a worker can initiate a pull cord or press a button to notify management, maintenance, or other individuals of a concern. The alert may also be activated automatically by production equipment when a specified condition is detected. Similar to a check engine light on a car dashboard, it's a method to notify others of problems in real-time so immediate action can be taken to contain or correct the problem before work continues normally.

Typical reasons for triggering an Andon signal can include identification of a defect, part shortage, equipment malfunction, or safety concerns. In these cases, work can be stopped or the problem contained, allowing work to continue until a solution is found. It is important to understand what caused the work stoppage by analyzing available data and evidence in an effort to implement continuous process improvements that prevent reoccurrence. Modern Andon systems can also include audio alarms, graphic displays, lighting systems, and text messages for notification.

Display Chart Preparation Tips

When making data visible, consider including the following items on a chart or graph:

- Date the chart was prepared.
- The person responsible for gathering, preparing, and displaying the data.
- Tolerances, targets, or reference lines.
- Date last updated and frequency of updates.
- A trend line.
- A sentence or brief paragraph summarizing what the chart is communicating.

In summary, available and easily understood data and information create a high degree of engagement and help focus people on the right priorities. If you are starting up a manufacturing operation or are interested in improving your current visual control practices, first define which KPIs are most important to monitor and manage operational performance. Consider the ones related to safety, quality, productivity, and costs as a starting point. Assign ownership for data gathering. Display data and monitor performance at a defined frequency, acting when necessary to maintain process stability and control when threatened by process abnormalities and deviations. Understand the source of data and information needed for preparing and presenting the desired process performance indicators when communicating to appropriate stakeholders. Schedule periodic meetings to review performance trends and manage any necessary actions accordingly.

Key Points

- Keep processes simple and visual.
- Visualization facilitates communication by displaying relevant information in a clear and understandable format to create awareness and trigger action when needed.
- Good visual controls facilitate the easy and quick detection of deviations from standards and target conditions.
- Data for Key Performance Indicators are gathered, visualized, monitored, and controlled.
- Visual control is about making systems and processes transparent and self-explanatory.
- Data and information transparency build trust.
- Data without a purpose, baseline or reference point is meaningless.
- Schedule periodic performance reviews to monitor short- and long-term trends.
- Tribal knowledge can be used to manage a good company, but facts and reliable data are also required to run a great company.
- Establish visual controls to highlight current operating conditions for monitoring and control.
- Visual controls help identify deviations from standards, allowing teams to recognize when the process is not working as designed.
- Effective visual controls highlight abnormal conditions for consideration and action.
- Audio and visual controls support many shop floor activities, including daily meetings, Gemba walks, layered process audits, and the execution of standard work routines. They display vital information for monitoring and control while facilitating transparency in operational performance.
- Communication boards are used to visualize KPIs, facilitating the search for process abnormalities and deviations from standards.
- Define how visual controls will be prepared, executed, monitored, and controlled in production as part of the manufacturing operations manual.

2.4

Shop Floor Communication

Speak of what you know and listen when you don't know.

Objective: Communicate to create awareness, understanding, and action.

DOI: 10.4324/b23307-10

Figure 2.4.1 Communication methods.

Overview

Communication involves the exchange of information between people. We communicate with each other in five different ways, verbally, non-verbally, in writing, through listening, and visually (see Figure 2.4.1). On the shop floor, we use all five of these approaches when managing operations. Employees talk with their peers and supervisors when discussing work tasks. Supervisors continuously monitor the non-verbal cues of employees to determine their comfort and competence in completing work activities. Documented procedures guide employees in completing their daily work routines. Management listens to employee input for awareness while all employees use their audio and visual senses to continuously monitor the process for abnormalities and deviations from operating standards.

Poor communication can slow or limit an organization's ability to achieve its tactical and strategic objectives in a timely and desirable manner. Communication failings can result in misunderstandings, operational inefficiencies, scrap, and rework, as well as stakeholder dissatisfaction. System optimization and operational excellence rely on accurate and timely communication practices.

Operational excellence does not happen without coordinated interaction between people working toward a common goal. Good communication facilitates collaboration and cooperation. When properly motivated, people can be encouraged to come together and discuss topics from which problems can be solved and decisions made. The more familiar and comfortable people are with each other, the more likely they will be open to conversing and sharing information and ideas that are essential for efficient and competent shop floor management. Sometimes, leadership simply needs to bring the right people together and get them talking and interacting on a regular basis. Operational success is the result of teamwork, not individual achievement. Teamwork occurs when people are effectively communicating.

Machine to Machine (M2M) Communication

Shop floor communication has traditionally occurred between people. This has been slowly changing with the evolution of technology. Machines are now communicating with people and between each other. Communication between machines is typically denoted as "machine to machine" or M2M. In this situation, devices can communicate wired or wirelessly. This type of communication is facilitated by application software using sensors and relays linked by a network. More recent developments have allowed these shop floor devices to communicate with people through mobile devices. Machines are now being configured to notify individuals of certain conditions or events when triggered during the manufacturing process.

Machine to Business (M2B) or Machine to Enterprise (M2E) Communication

Communication can also occur between the manufacturing floor and vendors that supply products and services to a facility. In this case, a communication link is established with an external source, allowing the selected supplier access to view shop floor data. In this instance, material consumption can be monitored or systems can trigger new orders when a low inventory threshold is reached or initiate a service request when a certain condition is realized.

Shop Floor Communication

Shop floor managers must be in constant communication with shop floor personnel, especially frontline workers, so they can remain up-to-date on current conditions and immediately react to circumstances that need their attention. To improve production floor communication among staff and frontline employees, consider the following:

- Participate in regular (daily) Gemba walks where the staff can review current performance data and engage in discussions with operators at the front lines of production.
- Maintain an "open-door" policy, allowing anyone to share their thoughts concerning operations.
- Facilitate open discussion with the staff to talk about issues, concerns, and problems.
- Remain visible and present on the shop floor. Work to build employee trust at all levels of the organization.
- Make a point to listen for understanding.
- Observe people's body language for signs of how they feel when expressing their thoughts.
- Be cordial and stay connected to staff members and workers through greetings and small talk.
- Encourage workers to learn new things and share their knowledge with those interested.
- Consider the latest mobile technology in employee communication. Available applications can establish real-time communication for issue notification and problem-solving.
- Provide frequent employee feedback so they know what they are doing is correct or have an opportunity to make timely adjustments accordingly.

It's important for managers and supervisors to meet with their peers and subordinates periodically to share information, discuss relevant issues, and evaluate manufacturing performance. Frequent, clear, broad, honest, and transparent communication, up and down the organizational hierarchy, is an essential attribute of effective shop floor communication. Let's take this opportunity to discuss some of the more common ways communication occurs on the shop floor.

Daily Meetings

Daily meetings, at all levels of operations, should facilitate timely sharing of data and information with those who need to know. Meetings can start at the production line with operators and their supervisors. Key topics from those discussions then move progressively up the hierarchy, as information is disseminated to create awareness, help solve problems, and make key decisions. These line-side meetings can occur more frequently, such as each shift, to maintain awareness, escalate high-impact issues, and drive necessary actions. Meetings that occur at the production line are typically brief, 10- to 15-minute stand-up meetings that are moderated by the line leader or area supervisor. Visual controls are often used as the basis for reviewing key process indicator (KPI) data on operational performance, discussing issues of concern and identifying opportunities for improvement. Topics are discussed with the primary objective of maintaining production line output targets.

A second- or next-level meeting is likely to occur shortly after the first, where team leaders or area managers raise issues that warrant upper management attention. Subsequent meetings should continue until the plant manager receives the data and information required to effectively run the plant. Along the chain of communication, data and information may be classified and routed to specific support functions (such as maintenance, quality, and facilities) who are expected to take ownership and act on information for which they have the responsibility, based on operational priorities. High-priority issues may need to be addressed immediately or escalated to the appropriate individual for a quick response.

Daily meetings are important but need to be focused and brief. These meetings should be structured and timely so the appropriate information is received by those who need it to identify the day's priorities and take action to address urgent issues. Employees must be prepared to share their information at the start of each meeting and provide a status of their assigned actions. See Figure 2.4.2 for an overview of daily factory meetings.

Weekly Meetings

Weekly meetings can be used to communicate and review performance trends of key metrics such as yield, scrap, rework, line output, and Over Equipment Effectiveness. Trend analysis provides an overview of operational effectiveness over the course of time and can help identify trouble areas, system weaknesses, and highlight improvement opportunities. With the advent of artificial intelligence, product and process performance data can be collected, analyzed, and highlighted to reduce defects and minimize equipment downtime when data is understood and acted upon in a meaningful way.

Weekly meetings can also provide an opportunity to review and discuss system-related topics such as equipment calibration status, tactical project progress, problem-solving activities, preventive maintenance schedule compliance, and environmental,

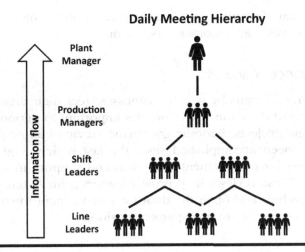

Figure 2.4.2 Daily meeting hierarchy.

health and safety issues. During the COVID-19 era, it was not unusual to discuss personnel and material supply issues on a daily basis due to the pandemic's impact on workforce availability and material delivery concerns throughout the supply chain.

Monthly Meetings

Monthly meetings are typically held with middle and upper management interested in reviewing overall plant performance. The topics of this meeting usually focus on functional activities and issues related to facilities, human resources, finance, logistics, and EHS (environmental, health, and safety), in addition to shop floor operations. These meetings tend to focus on performance trends, systemic action items, escalated issues, and strategic activities.

Facility review topics may touch on things like plant equipment upgrades, structural improvements, and energy usage. Human resource discussions can be about changes to fixed and variable labor, promotional opportunities for open positions, and diversity within the workforce. Monthly financial performance is usually considered from the perspective of actual results relative to budget and forecast of which significant deviations from these values are discussed. Determining the causes of significant variances in budget and forecast aids management in understanding the issues impacting performance and can help identify actions to offset short- and longer-term effects. Logistics will focus on the quality and cost aspects of the supply chain, from the purchasing of raw materials to the delivery of finished goods. The agendas for these monthly meetings could change from one meeting to the next, but they often target key topics to run the business.

Decisions and targeted actions resulting from these meetings are then communicated down the chain of responsibility for awareness, transparency, and follow-up. Since corporate sustainability has taken on increased interest and scrutiny over the past several years, this and other high-profile topics such as carbon neutrality and employee diversity, equity, and inclusion (DE&I) may become topics for discussion, if not already being reviewed. The primary scope of the monthly meetings is to assess compliance to the manufacturing system and ensure system outcomes are consistent with strategic objectives and customer expectations.

Project Performance Reviews

Strategic projects are a means by which businesses steer their organization in a different or more focused direction. Therefore, it's important to periodically review the progress of strategic projects. Periodic governance reviews allow project managers to share what has been accomplished since the last review, highlight issues and obstacles to progress for management awareness and support, in addition to outlining the next steps. These reviews help project leaders achieve timely completion of their project deliverables and open the door for management to coach and develop high-potential leaders for future management positions.

Gemba Walks

Gemba walks provide an opportunity to talk directly with people in the front lines of production. Frequent excursions to the shop floor promote face-to-face conversations with people having first-hand knowledge of what's happening in real time. Allowing employees to directly share their thoughts, ideas, and concerns with management reinforces their value to the organization, builds trust, and enhances morale. Gemba walks create shop floor awareness, facilitate the exchange of information, and allow management to understand problems and coach employees on the attitude and behaviors expected of the workforce. This type of interaction may also accelerate problem-solving when all the right people are in one place and talking.

CASE STUDY: GEMBA WALKS IN ACTION

Gemba walks are a great way to heighten management awareness of current shop floor activities, ongoing production issues, and performance results, in addition to enabling interaction between management and shop floor personnel. I have observed the effectiveness of Gemba walks in reestablishing operational stability at a facility that was disrupted by the transfer of over 25 production lines to its location after a plant, producing similar products within the company, was closed.

Due to the overwhelming influx of new equipment and the need to quickly qualify these "new" lines to meet customer demand, Gemba walks were started to understand the obstacles to process stability, capability, and control during production ramp-up. Participants in these Gemba walks asked questions of shop floor employees to understand the issues impacting output performance. Some of the fundamental questions asked were as follows:

- How many parts are we expected to ship this week from this production line?
- How many parts were produced yesterday relative to the plan?

- How many of these parts are currently in inventory?
- How many parts is this line expected to produce today during your shift?
- What issues with equipment, materials, people, or process are preventing the team from meeting its daily output goals today?

Teams are expected to take ownership of their line, understand what they needed to do to meet customer shipments, and allow them to vocalize any concerns they had during the Gemba walk to achieve their daily/weekly performance goals. Production line operators serve as the eyes and ears of management and are often the first line of defense in highlighting and preventing issues from impacting daily production and the ability to meet timely customer deliveries. In short, the right information, in the right hands, at the right time, can go a long way toward helping achieve and maintain stable and predictable shop floor operations.

Mobile Devices

Mobile devices have become an indispensable tool in modern-day manufacturing. They allow individuals to be in constant communication with manufacturing operations and the people who support production. Mobile devices can be used to access and receive information quickly, seamlessly, and remotely. Notification of emails received, upcoming meetings, equipment failures, and excessive scrap can all occur via mobile devices. Mobile devices, such as cell phones, tablets, and smart watches, can remind us to complete daily tasks, follow up on action items and respond to customer requests for information, data, or parts.

There is a category of mobile device that has been tailored for rough handling in a manufacturing environment. These devices are equipped with barcode scanners, dimensioning and thermal imaging cameras, wireless data capture, and sharing capabilities. As an example, bar code scanners can be used to monitor and control inventory levels, thermal imaging and sensors can facilitate machine diagnostics, and wireless vibration sensors can detect the start of motor bearing failures. When these devices are integrated into the internet of things (IoT) network, they can become immediate and valuable communication tools.

Productivity can improve with the application of mobile devices, as evident in the speed of capturing and accessing data for problem-solving and decision-making. These devices can record real-time data, making analysis and reporting more efficient. In certain cases, physical processes, such as paperwork and approvals, can be digitalized and completed via a mobile device.

In summary, good communication is key to good shop floor management. Great communication comes from the reliability and speed at which communication takes place. Digitalized communication moves organizations closer to a state of optimization as interactions with people and machines become easier, faster, and more efficient. On a more simplistic level, if we fail to communicate properly or do not communicate at all, we run the risk of confusing others. This can lead to misunderstandings, underperformance, and inefficiency. Thus, I will leave you with three words: communicate, communicate, and communicate!

Key Points

- Daily meetings typically occur once a day or shift to review operational performance and highlight key issues or concerns for escalation and action.
- Meetings are valuable when needed data and information are exchanged in a timely and efficient manner. However, meetings can quickly become distractions and non-value-added when they no longer focus on what is important. Be mindful of people's time, and be prepared for meetings.
- A communication plan helps ensure the flow of information to all levels of the organization happens quickly and efficiently.

2.5

Shop Floor People

You don't build a business; you build people and then people build the business.

~Zig Ziglar

Objective: Give people the knowledge, skills, tools, and freedom they need to do the job for which they were hired to do.

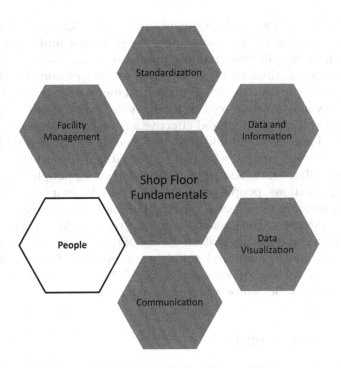

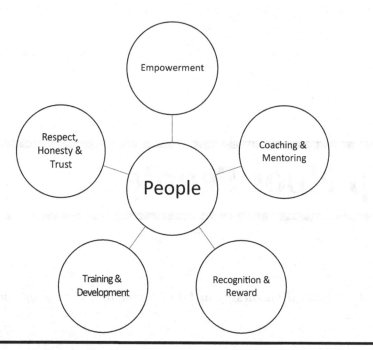

Figure 2.5.1 Treating People in the Workplace.

Overview

People are the heart and soul of an organization. They determine the rules, define the procedures, assemble the products, solve the problems, and make the decisions. People lead, sell, market, design, develop, purchase, receive, schedule, manufacture, audit, inspect, package, ship, and deliver goods to their customers. They do this as individuals and as coordinated teams brought together with specific goals and objectives in mind. They must work in concert to satisfy the expectations for which they are being compensated. How efficient and effective a team is, or can be, depends on how work is organized and executed and how well people are treated while doing their assigned tasks. In short, people who feel respected will respect their work and the others with whom they work. See Figure 2.5.1 for treating people in the workplace.

For many organizations, people are their most flexible, adaptable, and knowledgeable asset. Unfortunately, labor costs, especially in high-cost countries, can become a burden to competitiveness. Thus, many companies try to strike a balance between the use of people and technology. Regardless, the use of technology also comes with a price since highly skilled and experienced people are needed to develop, implement, and maintain operations. Let us explore some of the elements associated with the human workforce and how people can best serve a company and be served by the company.

Employee Empowerment

In today's knowledge economy, the human asset is the most valuable contributor to the profit, growth, and success of an enterprise. People are intangible assets that present creative ideas, develop products, execute processes, and deliver results expected to satisfy customers. People are considered intangible assets because an employee's skill set, expertise, knowledge, and experience cannot be measured in monetary terms.

Achieving manufacturing excellence requires companies to recognize their employees as assets that need to be nurtured and developed to continuously enhance company value. These assets bring valuable knowledge, expertise, abilities, skill sets, and experience to the work environment, setting the pace for growth and profitability. It is up to leadership to make sure that the work environment is ripe for maximizing the benefits employees can offer, especially when everyone is working toward a common set of goals and objectives aligned with the business strategy.

Companies distinguish themselves from their competitors, in part, by the human capital they attract and the environment they create to maximize the value these resources bring to the enterprise and their stakeholders. Employees are the company's face to customers and influence the perception stakeholders have of the enterprise. They are the source for the know-how and know-why required satisfy modern-day customer demands and industry challenges.

> People are definitely a company's greatest asset. It doesn't make any difference whether the product is cars or cosmetics. A company is only as good as the people it keeps.
>
> – Mary Kay Ash

Today, knowledge is more important than ever before. This emphasizes the importance of people and their contribution to organizations. It takes people to run a company. They shape the culture and influence the behaviors displayed within the workplace. They create the systems, define the processes, and write the procedures used to produce the products and deliver the services customers expect. The workforce must be committed and dedicated to following the principles, practices, systems, and procedures that constitute the framework upon which the enterprise is built. Poor attention to maintaining and strengthening the organizational foundation can lead to its rapid decline.

People have a tremendous capacity to change. They can learn and adapt quickly to develop new products, solve problems, make decisions, work in different teams, assist others, and act on their ideas. They can be firm and flexible, aggressive and calm, disciplined and imaginative, honest and demanding. Employees are one of the company's most important assets and greatest competitive advantage when empowered under the right working conditions. The key is to attract and retain the best talent possible, providing them with encouragement and a stimulating workplace while making them feel like a vital part of the company's mission.

Ultimately, companies are successful because their people are successful. This only happens when organizations make a conscious effort to create an empowering environment; an environment in which people can come together, apply their knowledge, and exercise their skills and know-how to effect change. We must empower people by motivating them with incentives, clear direction, and common goals while providing them with the methods and tools to accomplish their tasks and rewarding them when desired results are achieved.

Motivated employees drive change as they work to achieve performance targets, develop innovative products, and pursue new opportunities aligned with short- and long-term strategic objectives. Employees are the drivers of profit and loss. They influence customer perception and create a sense of brand loyalty. They do this through their understanding of how things work and the relationships they build with each other.

Employees are the backbone of the organization. They are far more important than machinery, production equipment, or the facilities within which they are housed. They have the ability to think, reason, solve problems, and make decisions, complemented by their ability to execute and validate decisions through action and results.

Skilled people are difficult to replace, especially if they have a deep and broad understanding of the company and its operations. It is often cheaper to keep experienced employees content than to hire new people with similar qualifications to replace those that have left. Retaining competent people should be the aim of every company. In addition, the continued development of high-potential people, who already know the company culture, systems, and procedures, may serve to enhance the benefits and value they already provide.

Respect, Honesty, and Trust

Respecting individuals is a core principle of operational excellence. Organizations should strive to hire and develop competent people who understand and can execute the work and advance the strategy of the business. Associates who have been hired for their motivation and capabilities should be provided direction and let go to do the work for which they were hired to do while providing support as needed. Decisions should be allowed to occur at the lowest possible level of the organization where the knowledge and experience exist to do so effectively. Trust people to make decisions within established boundaries while periodically confirming outcomes. Continuously mentor, coach, and provide support while holding individuals and teams accountable for expected results. People need to be directed, developed, empowered, and trusted to do what's required to achieve operational stability, optimization, and excellence. This requires all employees to be treated with respect and be provided with a clear purpose. That way, they can work effectively as individual contributors and as team members.

Effective employees are self-aware, self-organized, and work collaboratively with others for a common purpose. An effective organization is a learning organization that experiments to confirm what they know and learn what they don't. They apply their newly acquired knowledge to incrementally enhance enterprise performance over time. A learning culture also shares its knowledge with other stakeholders for the benefit of the entire organization. Disseminating lessons learned to suppliers, contractors, and customers alike can help elevate stakeholder performance and, in turn, can benefit all parties involved.

As the old cliché goes, honesty is the best policy. Honesty builds trust. People want to know the truth to better understand the world around them and to allow them to adjust their behaviors to seamlessly assimilate into the workforce. Honest communication and feedback help build relationships, facilitate cooperation, and strengthen commitment. Trust and mutual respect for people contribute to the foundation of a high-functioning, excellent organization.

Trust involves transparency and inclusion. It's about colleagues sharing valuable information with each other and across departments instead of using a "top-down" only approach. Employees should be encouraged to take ownership and be transparent about any difficulties or problems that arise. Elevating problems should be an expectation and treated as an opportunity to learn and continuously improve operations. Recognize and thank people for sharing their problems and concerns. Without this mindset, some problems never get solved.

Training and Development

Employees can contribute to a company's success by learning from experience, sharing their knowledge, and improving production efficiency. Learning, sharing, and improving are not difficult. It's a conscious and continuous effort every employee should practice in the workplace. It must become a workplace mindset and habit, especially since increased process and equipment complexity is demanding a higher level of know-how and know-why in today's work environment.

Industry 4.0 is the next frontier in manufacturing, focused on automation and data management. In one aspect, it's leading manufacturers to increase their application of robotics and data analytics in a production environment. This will stimulate new job expectations with higher cognitive capabilities and expanded skill sets expected of employees. Regardless of this trend, people will still be the driving force for change in the industry. As such, they will require continuous training and development to work with new technologies in preparation for future competitive challenges.

People need to know what is expected of them so they can adjust their behavior to align with those expectations. Developing people can involve classroom training, workshops, and on-the-job experience. This can be complemented by team leaders and supervisors mentoring and coaching employees on how to acquire new knowledge through experimentation. Experimentation can broaden one's knowledge base, lead to a deeper understanding of operations, and help to solve problems causing process instability and waste. Training and the development of people need to be commensurate with technological advancements and can be acquired through:

■ Sponsoring guest speakers on relevant topics.
■ Scheduling "brown bag" lunch lectures on existing and up-and-coming technologies.
■ Promoting membership in well-established industrial societies.
■ Attending key industry conferences.
■ Encouraging the pursuit of formal educational degrees.
■ Leading high-impact projects.
■ Reading and discussing recommended books, podcasts, and articles.

SIDEBAR: DEVELOPING PEOPLE THROUGH WORKSHOPS

An effective way to develop people is through workshops. Workshops combine training with the application of knowledge to realize tangible, meaningful results. When conducting learning/action workshops, qualified, capable, and motivated people can quickly and efficiently learn and apply their newly acquired know-how and know-why in the workplace. Workshops impart valuable knowledge, reinforced through real-time process improvements that demonstrate how data and information, shared with employees, can positively impact the work environment and operational performance.

Workshops are about engaging people in the act of practicing what they are learning. It is an opportunity for participants to go, see, observe, and question. It is about capturing meaningful data and information for review and discussion, which can be prioritized and put into action. It creates an opportunity to identify gaps between people's knowledge and understanding. It provides a chance for individuals to formulate questions, articulate disconnects, and bridge the gap between theory and reality.

A successful workshop requires an experienced coach, an individual who can lead participants through the process of learning and the application of knowledge to facilitate improvements. They lead workshop activities by outlining an approach, making suggestions, pointing out observations, and asking questions. Coaches are responsible for making sure workshop participants have the information and direction needed to be successful. Productivity improvement workshops should:

- ➤ Promote employee development.
- ➤ Facilitate learning, knowledge application, and confirm expected results.
- ➤ Identify opportunities for improvement.
- ➤ Elevate process maturity.
- ➤ Reinforce a mindset of continuous improvement.
- ➤ Minimize the need for additional investment.

Workshops can bring discipline to the activities of process control and continuous improvement by establishing a structured approach to eliminating problems and obstacles. They help participants learn and deploy a step-by-step methodology for achieving improvements. A workshop experience can be enhanced by a good facilitator or coach that works to keep the team focused on workshop priorities to achieve desired results with impact. As with any activity, it is important to have the full support of management, motivated team members, and a support network available to assist workshop participants when needed. Senior management should remain involved and engaged in workshops to mentor, coach, and encourage active participation.

Mentoring and Coaching

A company's on-going stability and manufacturing success is the result of a method-ical approach to identifying, mentoring, and coaching current and future leaders who can continuously execute their carefully developed tactical plans and long-term strategy. Mentorship involves advising and training high-potential employees to enhance their understanding and develop their reasoning skills. Coaching takes a similar role by providing employee training and guidance. Consider assigning men-tors to high-potential employees in order to evaluate their decision-making process and leadership capabilities. Tailored coaching can then be used to further develop their competences based on observed performance and expected behaviors.

Look to create a mentoring program where targeted employees are matched with willing leaders. These leaders can involve their mentees in key decisions, critical problem-solving, strategic reviews, and insightful meetings to expose, evolve, and refine their management skills in preparation for future leadership positions. As stated before, manufacturing excellence does not occur by chance. It takes a deep knowledge of an organization's strengths, weaknesses, opportunities, and threats to understand where an enterprise stands, and sufficient insight to recognize the right people within the workforce who can lead it forward to overcome the issues and obstacles that lie ahead.

Recognition and Reward

People want to be recognized for their good work. They want to feel valued and that their contribution to the company makes a difference. That is why employee recog-nition and reward programs are an important part of human resource management. A good human resource management system can help keep the workforce satisfied, motivated, and efficient while reinforcing desired behaviors. It's important for organi-zations to create an employee recognition and reward program that boosts employee retention, engagement, morale, and productivity. Allowing employees to recognize their peers can enhance such programs since management is not always aware of the good work exhibited by people under their employment.

Excellent companies should also recognize and reward people who collaborate across functional boundaries and within networks. The sharing of data, information, and ideas across a network of people can broaden the scope of opportunities for improvement, identify more potential options for problem-solving, and increase the number of ideas for consideration. When recognizing and rewarding people, strive to provide a reward of value that is consistent with the uniqueness of the individual. This approach will show a more genuine appreciation for the accomplishment. In addition, look to customize individual rewards to reflect their level of achievement. Completing a task vs. Leading a successful project demands a different response. Focus on what was achieved and its value to the company and its customers.

Recognition is most significant when done soon after a desired behavior or achievement is demonstrated. The longer you wait, the less likely the show of appre-ciation will positively impact an individual. Regardless, it is important to clearly and specifically state why recognition and reward are taking place. This not only reinforces the reason why; it also demonstrates to others what is considered acceptable and appreciated behavior. Finally, consider public recognition since people typically feel good when recognized for their achievements in front of their colleagues and friends.

Stakeholder Management

Dictionary.com defines a stakeholder as "a person or group that has an investment, share, or interest in something". In manufacturing, a stakeholder can be company's employees, management, customers, or suppliers. This definition may extend to other independent parties such as the government (when regulations apply) or the media if newsworthy topics are happening. In many instances, stakeholders should be managed by first understanding their role in the organization, their level of interest in certain activities, and their willingness to influence those activities in a meaningful way. Managing key stakeholders may require keeping them in the information loop, involving them in critical decisions, or engaging them in problem-solving. Regardless of the situation, stakeholders almost always exist and need to be assessed for their interest and potential impact on on-going operational activities.

Stakeholder management involves identifying the people, groups, or organizations that could impact or be impacted by manufacturing operations. To actively manage stakeholders, their expectations must be understood, assessed, and integrated into a plan that satisfies their needs in executing work and making shop floor decisions. Relevant topics for shop floor stakeholders may include data collection, analysis, control practices, issue escalation, personnel changes, process changes, performance results, and employee engagement in productivity improvements. It is important to keep stakeholders, most interested and affected by shop floor activities, up to date on current events and results. Consider various forms of communication to do so, including email, text messages, meeting minutes, and video conferences.

Keep contact direct and transparent, avoiding excessive communication layers, which tend to slow the information sharing process. Involve key stakeholders in critical meetings and decisions when appropriate. Regular interactions with stakeholders can help build a strong and cooperative working relationship. It can also avoid surprises, mitigate risk, and build trust.

In summary, leadership and management must embrace people, treat them with respect, train and develop them, mentor and coach them, recognize and reward them, and teach employees the ways of the company. They must provide a clear vision and direction, create a cooperative and collaborative working environment, and strive to motivate and incentivize employees to do their best work every day.

SIDEBAR: EMPLOYEE SATISFACTION

Employee satisfaction changes as different generations join the workforce and major world events shape people's perceptions of what is possible. In the wake of COVID-19 pandemic workers are looking for more remote work options, which require different technologies to support this desire for increased flexibility. There are many proven ways to keep employees satisfied, some of which are as follows:

- Benefits and perks.
- Incentives and bonuses.
- Workplace design.
- Recognition and rewards.
- Balanced workloads.
- Professional development opportunities.
- Open, transparent, and accepting cultures (DE&I).
- Dynamic and continuous learning environments.
- Work from home/anywhere options.
- Revenue sharing.

Key Points

- We don't change people with written words, we change them through setting the example and our daily actions.
- Make employee development an inherent part of the organizational strategy and culture with full management support.
- Prioritize people's health, safety, and satisfaction.
- Give employees the knowledge, skills, and tools they need for success.
- People need to feel their contribution matters, large or small. Periodically express your appreciation for people's efforts.
- Employees are responsible for the goods and services that deliver customer value. Happy, motivated, empowered and engaged employees translate into customer satisfaction.
- Hire and develop the right people. Allow them to acquire the knowledge, skills, and experience the company needs to maintain and enhance current operations while preparing for future challenges.
- Employees are the face of the company to customers. Treat them like you want them to treat your customers.
- Happy and valued employees drive profitable and competitive growth. Dissatisfied employees can quickly become a liability.
- A work environment of respect and trust creates an empowered workforce.
- Excellence is built on attracting the right people and developing them for managerial and leadership positions.
- People should be trusted until a point when they breach that trust.

2.6

Facilities Management

Facilities management is an integral part of successful shop floor management.

Objective: Maintain a stable operating environment that can support production needs.

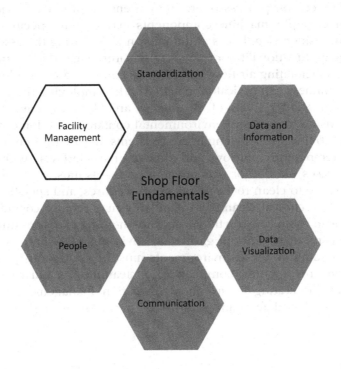

Shop Floor Environment

Manufacturing is made possible when the proper infrastructure is in place to support it. Facility management and support functions play a key role in providing a comfortable and suitable environment for daily work activities. Protection from the external environment, the uninterrupted flow of water, electricity, natural gas, and compressed air all play a part in maintaining a stable working environment and manufacturing operations. In addition to basic facility infrastructure, there are key factors to consider when working within a shop floor environment, including facility layout, workplace design, fire suppression, chemical storage, utility access, electrostatic discharge (ESD), and cleanliness, to name a few. These items should be considered when creating safe, stable, and compliant work conditions. In this chapter, we will review the impact of these factors on the working environment and consider the controls necessary to ensure they are properly managed. Let us start with cleanliness.

Technical Cleanliness

The cleanliness of a shop floor work area may go well beyond good housekeeping. A certain level of technical cleanliness may be required to avoid product defects caused by environmental contaminants such as dirt, dust, and human contaminants including skin oils, hair, and spittle. A measurable cleanliness level may be required for production or certain process operations to ensure product integrity and minimize scrap. For example, machine components may require specifications for particulate and film-residue cleanliness at the micron level during the assembly process to ensure adequate functionality or to avoid premature wear during application.

In addition to facilitating air flow and filtering controls, a certain level of product and process cleanliness can typically be achieved by employing the use of hairnets, coats, and gloves. Higher levels of cleanliness are likely to require stricter work procedures to ensure human and environmental contaminants don't impact product reliability or durability. Specially engineered systems may be needed to control work area humidity, temperature, airflow, and airborne particle levels within strict tolerances. In these cases, shop floor management protocols must be deployed to maintain strict adherence to clean room practices, procedures, and specifications.

Compliance to technical cleanliness standards may require considerable capital investment in clean room construction costs and on-going expenses due to employee training, purchasing of cleanroom garments, environmental controls, and constant maintenance. In addition, these environments must be continuously monitored and controlled to ensure on-going compliance to cleanliness standards. As with any closely controlled operating environment, deviation management becomes paramount to avoid product defects and scrap.

Ergonomics

Ergonomics, sometimes called human engineering, is the science of designing and arranging items people use to improve productivity and safety in the workplace.

It is a study of how people interact with their work environment. Manufacturing ergonomics focuses on the best way for people to handle materials and assemble products to minimize fatigue and discomfort and avoid injury. For example, a workstation can be studied for how an operator interacts with the equipment, tooling, and materials needed to execute their daily work activities. The results of such a study could lead to recommendations for changing workstation seating, lighting, and the rearrangement of equipment to optimize operator comfort and minimize their movement over an 8- to 12-hour work shift.

An ergonomic study focuses on five aspects: safety; comfort; ease of use; productivity (or performance); and aesthetics. The objective is to consider changes to tasks, workstations, tools, and equipment that minimize worker stress. This may include improving worker posture, regulating environmental temperature, and reducing body movements. Engaging an operator to periodically stretch and walk around may also be considered part of an individual's daily work routines to avoid a sedentary state that can result in discomfort or pain. Areas to consider when improving workplace ergonomics include:

■ Equipment layout and operation.
■ Material handling (lifting, pushing, and pulling).
■ Lighting.
■ Noise.
■ Task, job, and workplace design.
■ Workstation layout, and equipment height.

A more advanced look at ergonomics (e.g. cognitive ergonomics) extends into task analysis, where human behavior is studied in man–machine interaction. This knowledge can be used to design machine display interfaces and controls that are better suited to support workforce needs. When managing the shop floor, it is important to monitor worker comfort and respond to employee complaints. Complaints reveal opportunities for improvement.

Layered process audits can be employed as a way to monitor ergonomic conditions for continued effectiveness relative to changing operating conditions. Gemba walks can also be used to look for ergonomic improvement opportunities as workers perform their daily routines within their natural working environment. The following are some questions to ask when looking to improve assembly-line operator ergonomics:

✔ Observe operator movements and how they pick, handle, manipulate, and place material. Do they reach, bend, or move unnecessarily?
✔ Can you simplify the operator's head, hand, or feet movements?
✔ Observe how operators interact with their equipment; is it seamless and graceful, or do they struggle?
✔ Can fixture, tooling, or material placement be improved at the workstation? Can they be made simpler or easier to access, pick up, or engage?
✔ Do operators wait for long periods of time? Why?
✔ Review how the material is presented to or picked up by an operator. Can it be improved or simplified?
✔ Can the work area or workstation lighting be improved?

In the spirit of continuous improvement, constantly look for opportunities to improve operator work activities. They will appreciate the attention.

Electrostatic Discharge

Electrostatic discharge (ESD) is the sudden flow of electricity between two differently charged objects in contact. This event can sometimes be observed as a visible spark when a heavy electric field creates an ionized conductive channel in air. ESD can be an industrial safety hazard due to its potential to explode gas, fuel vapor, and coal dust. When sensitive electronic devices are being handled in a manufacturing facility, such as solid-state electronics and integrated circuits, precautions must be taken to prevent damage from an unexpected ESD. In these facilities, efforts are made to create electrostatic discharge protected areas (EPAs) through the use of ESD-safe packing materials, assembly worker clothing (with conductive filaments), conductive wrist straps and foot-straps, antistatic mats, and flooring materials. These are used to conduct harmful electric charges away from protected work areas. ESD-sensitive devices will also need protection during shipping, handling, and storage. Static buildup and discharge can be minimized by controlling the surface resistance and volume resistivity of packaging materials.

The objective is to remove static at workstations by grounding all conductive and dissipative materials and human workers by providing antistatic devices and controlling humidity. Humid conditions prevent electrostatic charge because the thin layer of moisture that accumulates on surfaces will help dissipate electric charges. For more information on international standards for typical EPAs, refer to the international electrotechnical commission (IEC) or the American national standards institute (ANSI). There is much to consider beyond the obvious shop floor activities.

Utilities

Reliable utilities are essential for many manufacturing facilities since the abrupt or long-term loss of power, water, or fuel can have a significant impact on operations. The sudden loss of power can cause equipment to stop in a state that requires substantial time and effort for restart and production ramp-up. Both natural and man-made events can cause utility outages that disrupt operating conditions unexpectedly.

The best defense against these unforeseen events is to have recovery plans in place to address the major disruptions that are likely to occur. For example, the supply of energy may not be as reliable in some countries as it is in others. In these cases, backup generators and independent energy sources (e.g. wind and solar) may alleviate the impact of these types of outages. An early warning system of an impending event may allow a facility to shut down key equipment and systems, and reinforce needed structures at whatever time is available. Start-up procedures should be available for equipment that may need more time and attention than normal to restart. These are only a few of the many ways to prepare and quickly respond to unexpected events, to resume normal operations as soon as possible. Time to recover can be significantly reduced when plans are ready for execution in response to unexpected yet not necessarily unlikely events. Remember, an ounce of prevention is worth a pound of cure.

Key Points

- Facilities can play a key role in the quality and reliability of processes to produce consistently good products.
- A high level of technical cleanliness may demand a high level of facility attention and budgeted costs.
- Ergonomics helps to address the human side of shop floor management.
- Be prepared for the unexpected and be ready to respond accordingly.
- Reference ESD and EPA standards to obtain a better understanding of how to properly control electrostatic sensitive areas in manufacturing.

Source

[1] https://www.sciencedirect.com/topics/engineering/cognitive-ergonomics

SHOP FLOOR MANAGEMENT

> *Manufacturing is more than just putting parts together. It's coming up with ideas, testing principles and perfecting the engineering, as well as final assembly.*
>
> ~ James Dyson

Objective: Maintain a stable, capable, and efficient manufacturing operation with engaged employees.

Overview

To create an efficient and effective operating environment, an organization must establish structure, discipline, and accountability in the workplace. Organizational structure provides awareness and boundaries within which people are expected to work and thrive. The degree to which structure exists may vary based on the size, complexity, and operational rigor of an enterprise. Highly structured organizations typically articulate guiding policies, principles, and practices to create an intentional work culture. They define a strategy to align employees with a common vision and mission, define standards and procedures to minimize process variation, and work to achieve stable, capable, and controlled processes. They also seek to continuously improve operational efficiency. Less structure equates to less influence and predictability over process activities and outcomes. Little to no structure is an invitation for chaos.

Managed chaos occurs when daily work activities are dictated by unexpected shop floor events creating a reactive approach to management. More structure and discipline are typically required when seeking a proactive and predictable manufacturing environment where daily routines are planned and executed versus prioritized by the latest "fire" needing to be extinguished. A structured approach to shop floor management should consist of defined process control activities in combination with

DOI: 10.4324/b23307-13

periodic productivity improvement actions. These are likely to be complemented by deviation management, change management, Kaizen events, and Jishuken workshops. Proactive management requires a continuous and real-time awareness of what's happening on the shop floor and the knowledge to take appropriate action in response to process deviations and abnormalities. Knowledge is a driving force behind making good decisions and permanently solving problems. Reliable data and information facilitates informed decision-making which leads to meaningful actions needed to achieve desired results.

Part 1 outlined the foundational elements for shop floor management. Part 2 explored shop floor fundamentals required to create a robust foundation for manufacturing operations. In Part 3, we focus on the daily shop floor activities required to achieve and maintain process stability, capability, and control. We will start section 3 by discussing standard work routines (SWRs) required to effectively execute the shop floor management system. SWRs are created by the workforce to articulate and prioritize the periodic work tasks necessary to proactively maintain operational control over the process and react to abnormalities whenever detected during daily monitoring. Process execution, monitoring, and control are essential shop floor management activities.

Daily shop floor management (DSFM) is the oil that keeps the engine of operations producing output at a predictable level of performance. If there is not a clearly defined or executed shop floor management system, process stability is temporary at best, with underperformance a common occurrence. If you don't recognize and understand the importance of DSFM in maintaining stable, capable, and controlled operations, you will never be able to optimize operational performance. This makes daily shop floor management an essential practice for any manufacturing facility seeking stable, consistent, and predictable output results. When DSFM is complemented with organizational structure, operational discipline, and employee accountability, a robust manufacturing system can be created to maintain operational control, drive sustainable improvements, and realize excellence with the agility and efficiency demanded by today's competitive work environment.

Manufacturing culture is an integral part of daily shop floor management. It reflects how people behave and act in response to their work environment. Building, maintaining, or changing a manufacturing culture requires planning, execution, monitoring, and control: lots of control. **Planning** is reflected in the organizational structure (see Figure 3.0.1) defined and deployed by management to articulate the behaviors and actions expected of employees (Parts 1 and 2 of this text). It creates a framework within which people are expected to work. **Execution, monitoring, and control** are defined by the daily shop floor tasks associated with individual roles, responsibilities, and work routines (Part 3). These activities need to be standardized, deployed, and continuously reinforced by management. This paves the way for sustainable productivity improvements built upon a structured and disciplined operating environment. In short, robust manufacturing is rooted in a sound foundation of enterprise culture, organizational structure, operational discipline, and employee accountability with success being defined by the results of work performed. Operational performance reflects how well the management system is

Figure 3.0.1 Fundamental Elements of Daily Shop Floor Management.

planned, executed, monitored, controlled, and continuously improved. Let's explore the world of daily shop floor management in the next chapters and discuss five key elements that influence effective deployment of the management system: man, machine, methods, materials, and the environment.

Key Points

- In part, shop floor management can be viewed as a continuous cycle of activities including process planning, execution, monitoring, and control.
- As the manufacturing management system is executed, data and information are generated, reviewed for abnormalities, and system adjustments made accordingly.
- Improvement opportunities are continuously identified for consideration, disposition, and implementation, upon approval.
- The essence of daily shop floor management is maintaining and improving process performance.

3.1

Manufacturing System Execution

Plans need to be executed to have meaning and deliver results.

Objective: Execution of the daily shop floor management system requires action. No action, no result, no change.

Overview

Once the management system has been defined and documented (e.g. planned) in a manufacturing operations manual or guideline (Part 1) and the foundational elements of the operating system have been put in place (Part 2), it's time to execute the manufacturing system as designed. Execution is an ongoing activity that will benefit from a clearly defined set of work routines assigned to key functions and individuals within operations. These routines establish the frequency at which repetitive tasks are performed on the shop floor to maintain system integrity through process monitoring and control. A high-volume manufacturing operation requires the identification and execution of standard work routines to ensure stable and capable processes continue to produce predictable results that consistently meet stakeholder requirements. It's similar to visiting your favorite restaurant, you return each time with the expectation of good food and service. This can only be achieved if the work routines used to prepare and serve your food are understood and followed, regardless of staff turnover.

Standard work routines capture "tribal" knowledge and create a roadmap for executing work within operations. These same routines serve as the conduit for consistent and predictable output performance. Successful operations rely on documenting the best known and most efficient way to perform work and the value of executing work as documented, every time, until a better way is discovered. Operational excellence can only be achieved when the way of working, required to realize excellence, is identified, understood, repeated, and continuously improved.

In this chapter, we will explore the role of standard work routines in managing shop floor activities. Although there are many different activities required to manage the shop floor, a properly defined, allocated, and executed set of routines will go a long way to reducing operational complexity by clearly defining functional tasks, task ownership, and expectations.

Standard Work Routines

Standard work routines, sometimes called leaders standard work, are a set of work tasks for individuals to execute, at specified frequencies, in order to maintain system integrity and operational performance. They define who, what, when, where, and sometimes why it is necessary to complete high-value work activities. These activities support the fundamental workings of the manufacturing management system to maintain its balance, robustness, and effectiveness.

Work routines should be focused on maintaining the most efficient operation possible through the continuous monitoring of process inputs and activities, addressing deviations from standards and targets, and looking for constant opportunities to improve productivity in a never-ending cycle of Plan-Do-Check-Act. Work routines are defined for the roles in each function or department that support the shop floor. Think of them as the third element within the roles, responsibilities, and work routines pyramid (see Figure 3.1.1). At this level of standardization, daily repetitive work tasks and activities are being defined at a given frequency (every shift, daily, weekly, monthly) to maintain and enhance operational performance. They should detail the work required for maintaining process stability and capability to the degree that if someone leaves the company, another person can seamlessly step in the role and do most of the work by executing the standard work routines already defined for the function. Tribal knowledge does not need to walk out the door with people exiting the company.

It's common knowledge that all tasks are not created equal, some being more important than others. When executing daily work routines, priority should be given to the most important activities of the day, knowing that unexpected events may

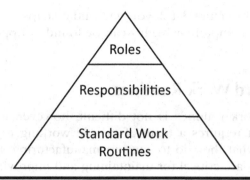

Figure 3.1.1 Roles, responsibilities, and standard work routines hierarchy.

draw one's attention away from completing all scheduled routines. In doing so, you ensure priority tasks are completed first in anticipation of unexpected interruptions. Safety- and compliance-related tasks need to be prioritized. Contractors and visitors to the plant should always be aware of these expectations and demonstrate compliance. Process and procedural deviations should not be tolerated with appropriate action taken to correct abnormalities. This requires that deviations be understood for root cause and countermeasures taken when needed.

When first preparing standard work routines for deployment, focus on eliminating current employee tasks that don't add value to daily operations and look to eliminate the obstacles to completing routine tasks that will add value. Routines should not define 100% of an individual's work activities since time is needed each day to work improvement projects and address unexpected issues that frequently occur in manufacturing. Keep standard work routines to no more than ~80% of an individual's workday capacity.

One key to preparing a set of standard work routines is to know what key process indicators to monitor, when to execute the routines, and when to react to process abnormalities, based on an understanding of their root cause. In addition, there is no one "right way" to do a routine or task. Identify the safest, cheapest, fastest, and easiest way to do it and do it that way every time while always look for a better way to doing it. Functions supporting operations can also be assigned daily routines based on an assessment of their repetitive work tasks. Regardless of what people think or say, there is most certainly a percentage of everyone's manufacturing work that is repetitive, although the percentage may vary among functions. For example, quality should be reviewing scrap frequently and working on the top failure causes for corrective and preventive action. Supervisors should be checking operators periodically to verify their compliance to standard work instructions. Management should be leading regular Gemba walks and layered process audits on the manufacturing floor to elevate their awareness, coach employees, and reinforce desired behaviors. Almost everyone has some repetitive work routines. They simply need to stop and think about what work tasks they should do every day, every week, and every month to support manufacturing operations. These become their standard work routines. It's not that hard to do nor difficult to document yet, people may be surprised by what they discover. What may be hard for some people to do is exercise their routines, every day, as defined. In Figure 3.1.2, you will find a snapshot example of standard work routines. A more comprehensive list can be found in Appendix 3.

Preparing Standard Work Routines

Preparing standard work routines is not difficult; however, it does take time and some thought. First, it requires a team of people working together in production to meet and discuss what they do to support manufacturing. From that discussion, they decide what tasks are critical for maintaining and improving the manufacturing system. They must also decide which current activities are non-value-added and constitute waste. The individual team members then document the essential work activities, for their role and responsibilities, necessary to maintain daily operations within a state of control. Once ownership for the critical work routines are established, they are prioritized, a brief description of each is documented, and frequency of execution assigned. The remaining work, classified as non-value-added, is then targeted for elimination.

The following steps articulate the actions to convert work activities into standard work routines:

- Identify a functional group with manufacturing responsibilities.
- Schedule a meeting to discuss their role, responsibilities, and activities in port of manufacturing.
- As a group, define or outline the *essential tasks* required to support manufacturing operations, independent of what's currently done.
- Compare these essential tasks to the ones currently being done, refine the list.
- Prioritize the activities, with the most important ones clearly highlighted.
- Identify all non-value-added tasks and target them for elimination.
- Provide a brief description for each standard work task and define a corresponding frequency of execution.
- Finalize the list of standard work routines, perhaps in the form of a procedure, work instruction, or checklist.
- Start executing the standard work routines, at the defined frequencies. Expect to update and change the routines based on lessons learned and experience with the initial roll-out.
- Continue to update/refine the routines as manufacturing activities evolve over time.
- Update the manufacturing management system to perform periodic checks to verify proper and continuous execution of the standard working routines. Reinforce and modify behaviors observed accordingly.

It's important to prioritize activities on the standard work routine list so it's clear what must get done each day, each week, and each month to ensure system integrity. The expectation is that high-priority actions get done on schedule, while lower priority tasks that don't get done on time, every time, will not significantly impact process performance, unless they are continuously neglected. If you complete the most critical activities first, they are more likely to get done, regardless of the disruptions

| STANDARD WORK ROUTINES | | | | | | | Suggested Starting Frequency | | | | | | | | |
Activity	Reference Documents	Checklist / Agenda	Line	Plant	Description of Routines	Shift	Daily	2X/W	W	2X/M	M	Q	A	Action / Notes
5S Method		Checklist	X		5S Compliance - Perform a quick 5S compliance check of the production Line	LL	S							PM = Plant Manager
				X	5S compliance - Perform a 5S assessment (shop floor, warehouse, support operations, administration, etc...)				S	FM	PM			FM = Factor Manager
					Performance Review - Review 5S functionality & effectiveness; review 5S scores for stability & trends					All	All			Q = Quality
					5S Roadmap - Review execution of plant 5S roadmap to plan. Address any deviations.					All	All			S = Supervisor
Continuous Improvement				X	CI Activities - Perform a random review of plant CI activities (e.g. Kaizen events, Jishuken workshops, etc.)				S					LL= Line Leader
		Process Maturity Assess	X		Process Maturity Assessment - Confirm a line maturity assessment was performed & improvement actions are inprocess. Evaluate effectiveness of completed & current activities. Coach as required.					S	FM	PM		
		Agenda			Performance Review - Review status of plant Improvement projects						All			
EHS			X		Confirm safety policy and procedures are being followed.	LL								
ESD			X		ESD Line Compliance - Confirm ESD line compliance	LL			S					
					ESD Compliance Review - Review ESD Audit results (plant level)						All			
Deviation Management			X		Line DM - Deviations from standards and procedures (e.g. SW, LSW, KPIs, etc.) are being identified, properly addressed and closed in a timely manner	LL	S							
			X		Line Confirmation - Confirm DM is performed, functional & effective. (e.g. SW verification, 5S, Jidoka, KPIs)		S		FM		PM			
					DM Plant Confirmation - Confirm deviations are found & addressed by plant staff (Gemba Walks, LPAs, Performance Reviews, etc.)			FM	PM					
Gemba Walk		Yes	X		Perform Gemba walk of production line		S							
				X	Perform general or area specific Plant Gemba walks									
					Confirm gemba walks are efficient, effective & actions identified are prioritized & followed-up.						All			

Figure 3.1.2 Snapshot of standard work routines.

often encountered on the factory floor. In addition, it may not be unusual to find a lot of non-value-added work that was once important but no longer relevant. If additional tasks are identified during the process of developing standard work routines, it becomes even more important to eliminate any work that is not positively contributing to bottom line results.

Key Points

- Process output results reflect the quality and control of manufacturing inputs and activities.
- In its most fundamental form, standard work routines prioritize the periodic work activities people should do to maintain process control. It's a "checklist" of essential activities needed to maintain daily operations and manage deviations from operating standards and targets.
- Standard work routines extend beyond employee roles and responsibilities by defining the specific and repetitive daily, weekly, and monthly tasks required to keep the operating system "in check" and performing as expected.
- Standard work routines involve monitoring and controlling key aspects of the process.
- Standard work routines provide structure that helps leaders focus on the daily priorities for efficient shop floor management.
- Standard work routines indicate what employees should do and eliminate the activities they should not.
- Standard work routines define the periodic activities expected of manufacturing support functions; it's their daily work instructions.
- The manufacturing system must ensure employee adherence to their standard work routines as a form of "checks and balances".

3.2
Manufacturing System Monitoring

If you focus on principles, you empower everyone who understands those principles to act without constant monitoring, evaluating and correcting, or controlling.

~ Stephen Covey

Objective: Continuously observe shop floor activities, data, and information looking for process abnormalities and deviations from standards and targets. Use this knowledge to prioritize actions for process control and continuous improvement.

Overview

Manufacturing system monitoring is about observing the process, checking for system compliance to standards and procedures, and verifying the quality of work performed meets expectations. It's a continuous review of key performance indicators and output performance from which insight and analysis are used to facilitate system control and identify improvement opportunities. Monitoring is an on-going process involving collecting, measuring, and analyzing performance data. It involves assessing performance activities and trends to trigger action required to maintain process control and drive efficiency improvements.

Monitoring helps ensure process system inputs, activities, and corresponding outputs are correct, complete, and meet operating standards. Work performance data and information generated from system execution are monitored for process abnormalities and deviations which, when discovered, will drive action to maintain system control. System control is the topic of the next chapter.

A shop floor management system is not one system, it's many systems working in concert to maintain process stability and produce consistent, predictable output. Systems can be complex so keeping them working efficiently together is an on-going activity requiring time and effort. To do so effectively, systems need to be continuously monitored so that abnormalities can be detected, analyzed, and addressed. Shop floor management requires system monitoring through observation, audits, and periodic reviews, to verify system effectiveness, product quality, and adherence to operational standards.

If you are constantly firefighting, you have lost control over your processes and are likely being influenced by them, not actively managing them. In this case, you may need to define more structure and exert more discipline in executing daily work routines. This includes increasing the frequency at which processes are checked for adequacy, proper deployment, and effective follow-up with timely corrective action when deviations are identified, to avoid generating defects. Once stability is established or reestablished, process monitoring, as a trigger for control, must be an on-going activity of operations to ensure system stability over time. Typical system monitoring activities include:

- Going to the source of work to observe, talk with people, collect data, and gather information.
- Reviewing data and information for abnormalities and deviations from standards and performance targets.
- Suggesting corrective or improvement actions for follow-up.
- Escalating issues of concern for organizational awareness and management support.
- Coaching and mentoring people on proper system management.
- Making recommendations for next steps.
- Looking for process improvements opportunities.

There are many ways to monitor a process. Some of the more common ways include shop floor observation, Gemba walks, process audits, performance reviews, and data analysis. In the following chapters, we will review various ways to engage in shop floor monitoring with the understand that it's an essential practice that requires structure, discipline, and accountability to do it right.

Process Monitoring/Verification

Manufacturing systems need to be actively managed to remain efficient and effective. Managing a system often requires verification of proper system execution, continuous monitoring for abnormalities and taking corrective action to prevent product or service defects. Manufacturing system execution activities can be defined by standard work routines, as discussed in the previous chapter. Monitoring the completion and correctness of daily shop floor tasks helps prevent system degradation. The frequency of performing verifications is typically documented in a function's standard work routines. Monitoring standard work routine execution is one of the several critical activities for managing the shop floor. Other topics requiring attention include the monitoring of production line output performance, work instruction deployment, equipment maintenance, key performance indicator (KPI) trends, and problem-solving progress. The following practices are ways to effectively monitor manufacturing system performance and should be part of the standard work routines exercised by a well-organized and operated facility. Let's start the discussion of shop floor monitoring and verification with some of these fundamental practices. See Figure 3.2.1 for an overview.

Go, See, Observe

The objective of go, see, and observe is simple. Take time to periodically go and informally observe the processes over which you have responsibility. Stop and see what's happening. Note anything unusual. Ask questions of employees for clarification. The more often you spend time on the production floor observing, the more you are likely to see. Shop floor management is primarily accomplished through communication and observation, but nothing beats going to the hub of activity to see and understand what's going on in the facility first-hand. You can prevent unexpected problems by visiting the floor and talking with employees, but true excellence is going a step further by spending time looking for potential problems to solve before they become disruptions or worse, produce defects that escape to the customer.

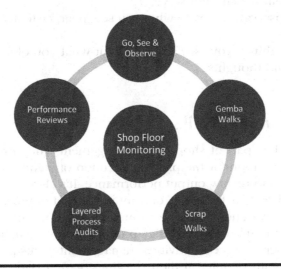

Figure 3.2.1   Shop floor monitoring practices.

I understand that stopping and seeing what's happening on the shop floor takes time, but it's often time well spent as you build a rapport with employees and work with them to make their jobs easier, faster, safer, and, sometimes cheaper, while looking for opportunities to improve operational efficiency. One of the best applications of the "go, see, and observe" mentality is during problem-solving. Going to the source of a problem creates understand and often provides insights that "armchair" problem-solving can't do. We observe to gain information, to learn, to understand. If you are interested in improving your observational skill during shop floor walks, consider the following:

- Know your subject. The more you understand, the more you are likely to notice.
- Slow down, stop, and look around. Focus your attention on your surroundings.
- Notice as many details as possible. Call out the details in your head.
- Use all five senses (see, touch, hear, smell, and taste) to experience the world around you.
- Walk with another person or an expert. Ask them what they see.
- Look at things from different viewpoints. What you see may depend on where you stand.
- Periodically change your focus from people, to objects, to surroundings, to things in motion, to the source of sounds.
- Look up, look down, look straight, look around.
- Look for patterns in people's activities and objects in motion.
- Maintain your observational focus. Don't allow your mind to drift.
- Actively listen. What do you hear? What sounds are unusual or out of place?
- Be curious. Ask questions.
- Try something new. Make observations at different times or under different conditions.
- Take a still photograph or video and review it. What do you observe?
- Cut out distractions to improve your concentration. Turn off your phone or find a quiet spot from which to observe.
- Record your observations. Note what you see, hear, taste, feel, and smell. Try to be specific.
- Go beyond the things you "see". Think about what you observed. Analyze your observations and thoughts.

Gemba (Management) Walks

Gemba walks are a key part of shop floor management. They provide management with an opportunity to monitor the proper execution of standard work instructions, daily work routines, and review output performance. It's also a way to slowly change organizational mindset and behavior, one interaction at a time. Gemba walks are conducted to create awareness, enhance communication, monitor for problems, and respond first-hand to issues detected. They provide prompt management and leadership to routinely observe process activities. Management walks provide plant staff an opportunity to talk with people on the front lines, gather data and information for understanding, engage in problem-solving discussions, coach decision makers, and identify opportunities for continuous improvement.

During a Gemba walk, look for obsolete information, negative KPI trends, redundant activities, excessive data collection, the need for additional training and misalignment of employee behaviors with organizational expectations. Strive to address issues encountered immediately. If unable to do so, follow up quickly with an action item or assignment of a problem-solving team. Define responsibility for action items and drive timely closure.

Engaging in daily Gemba walks is a commitment by management to maintain operational awareness, build employee trust, and reinforce the disciplines for process control and productivity improvement. It presents an opportunity to reconcile what's happening relative to what's being said. Gemba walk participants can provide support in addressing process deviations and offer an opportunity to coach and lead teams through systematic problem-solving.

Gemba walks are not a one-way interaction. It's a collaborative gathering of shop floor personnel with the management team. It's about changing the working relationship of company employees; creating a participative and collaborative work environment. All parties and their contribution are equally relevant. Gemba walks are as much a relational gathering as they are performance reviews. Team cooperation, personal relationships, and operational performance are critical for optimal outcomes.

When organizing or preparing Gemba walks, consider inviting functional managers or their representatives, if they can't attend. Gemba walk participants usually include the area supervisor, Maintenance, Production, Quality, Engineering, and Logistics team members. However, feel free to invite anyone who can add value to the Gemba walk experience. A typical Gemba walk agenda or checklist may look like the following:

Gemba Walk Checklist

- **Production plan** – is a build schedule available for the week and are there any issues of concerns in meeting the schedule when executed?
- **Production output** – was recent output performance consistent with the build schedule? If not, why not?
- **Deviations** - were significant issues or obstacles impacting production highlighted and addressed?
- **4Ms** – are there any new or open issues related to Man, Machine, Materials or Methods?
- **KPIs** - are KPI trends within targets?
- **Cleanliness** - is the area 5S compliant?
- **Process Verification** – is operator work periodically being verified as compliant to area standard work instructions?
- **Action Items** – are there any due or overdue action items to discuss? Note: every action should have a due date.

There are always opportunities to improve performance. Gemba walks are no exception. As such, I am including some Gemba walk tips for consideration. Feel free to use them, add your own tips from lessons learn, and share them with others as you teach, mentor, coach, and interact with company associates.

Gemba Walk Tips

- Schedule daily shop floor Gemba walks at the same time every day so people can prepare for the reviews.
- Make Gemba walks part of employee standard work routines.
- Gemba walks should be both structured (e.g. set time and agenda) and unstructured (e.g. casual walk through the shop floor).
- Make time for a personal Gemba walk (unstructured). Occasionally walk the production floor, stop, look, listen, and ask questions. Engage with shop floor employees.
- If you see something is not right, correct it immediately. Do not walk by it without acting, or at the very least, making a commitment to follow up.
- Don't assume and verify. Gather information and data to heighten awareness, solve problems, and make decisions.
- Review closed action items for effectiveness and permanent resolution. Take additional action, if needed.
- Teach Gemba walk participants to see through different eyes!
  - *Seek out problems* – look for deviations by comparing actual performance to standards and targets!
  - *Look for waste* – question why certain things are done.
  - *Do no harm* – don't berate people, develop them. Ask questions respectfully!
  - *Show respect* – builds relationships and trust!
  - *Make improvements* – look for opportunities to make immediate improvements to the area.
- Visually and verbally recognize people's good work in well-performing and improving production lines.
- If you can't visit every production line daily, make a point to do so over several days or within a week.
- Make yourself visible on the shop floor, people don't like talking to strangers.
- Ten minutes on the production floor is better than zero minutes.

SIDEBAR: THE 10-MINUTE GEMBA WALK

Step 1: Go to the Gemba
Start the day or afternoon by going to the Gemba. True leaders lead from a position of knowledge and strength. They go to the Gemba to increase their awareness, communicate, interact with people, and build relationships with their employees.

Step 2: Observe and Collect Data
Assess the current state. Take a few minutes to review available data and information. Look for abnormalities or deviations from process standards and targets. Collect information by talking to people.

Step 3: Discuss Key Issues with the Leadership Team
Review critical problems or concerns with responsible individuals and the leadership team. Agree on necessary actions through consensus. Make necessary decisions.

Step 4: Act
Implement actions, engage in problem-solving, discuss continuous improvement activities. Define follow-up actions with due dates.

Step 5: Follow up, Close and Confirm
Start the next Gemba walk by following up on open issues based on production priorities. Ensure issues are closed and confirm effectiveness of corrective actions or process improvements.

Scrap Walks/Reviews

Scrap is the Achilles heel of manufacturing. It must be closely managed since scrap directly and negatively impacts bottom-line profitability. The cost of material, labor, and time managing non-conforming materials all contribute to waste in manufacturing. To minimize scrap, it must be monitored and managed quickly and efficiently, so it does not continue to detract from process performance. The shop floor attitude toward scrap must be zero defects! To achieve this objective, scrap reviews should be performed frequently. To eliminate scrap you must identify when it occurs, understand why it occurs, and take action to prevent reoccurrence.

To manage scrap on the shop floor, start by scheduling regular scrap walks with Quality and Engineering. A daily review of suspect and defective parts allows the team to review commonly occurring issues and target process weaknesses for action. Scrap walks should consider the type and quantity of scrap accumulated during a shift, or specified production, period to prioritize next steps. Suspect and scrapped units provide a window into the cause of non-conformances and can help to implement countermeasures that prevent reoccurrence. In certain cases, a deep analysis may be required to discover root cause and implement containment, corrective or preventive actions. The responsibility for deep analysis may also be assigned to a supplier suspected of being the source of a problem or root cause.

When conducting scrap walks on the manufacturing floor, take the opportunity to talk directly with line leaders and operators to understand conditions under which various abnormalities and defects occurred. It's good practice to create a pareto chart of defects to assess the impact and cost of these issues on manufacturing performance. Visualized data can help direct problem ownership to the responsible party for resolution. To ensure stable manufacturing performance over the long-term, clear targets for scrap reduction should be established. By investigating and analyzing all rejected products, incremental actions can be taken to permanently eliminate defects from reoccurring, improving product quality, and manufacturing productivity. The following actions are recommended for managing shop floor scrap.

Actions for Managing Shop Floor Scrap

- Review scrap every day or every shift.
- Place questionable product on hold for review and disposition.
- Handle scrap appropriately. Clearly mark and separate scrap from good product and normal workflow.

- Stop the process if scrap exceeds an unexpected amount or specified limit.
- Disposition every unit scrapped for cause.
- Perform failure analysis for the cause of scrap, if unknown.
- Prepare a pareto chart of scrap causes to prioritize scrap reduction efforts.
- Classify product scrap as internal or supplier responsible.
- Pursue root cause corrective action.
- If not immediately possible to eliminate root cause, implement a containment before restarting production.
- Follow up on open actions.
- Monitor "percent scrap" as a KPI. Look for a trend in scrap reduction to validate effectiveness of corrective actions.
- Consider implementing Jidoka.

Layered Process Audits

Another method for monitoring and confirming manufacturing system compliance is through layered process audits (LPAs). LPAs are periodic audits performed by various hierarchical levels or "layers" of management to confirm manufacturing system functionality and effectiveness. Audits are conducted to ensure the methods or subsystems in place are working as intended and are producing desired results. Audits can help reduce process variation and defects through repeat verification that operators and other key functions are following standardized processes. They can be performed on work standards, training classes, projects, and process operations to name a few. The intent of these audits is to maintain an expected level of manufacturing performance.

LPAs can provide a way to systematically assess adherence to standards, processes, and procedures. They can be used proactively, as a risk management tool, to identify issues that may impact process performance downstream from the workstation being assessed. They can also be structured to verify error-proofing activity effectiveness. A well-managed LPA system will help identify deviations in key process activities that could be swiftly addressed to prevent defect generation. A properly targeted process audit can lead to a reduction of internal scrap, rework, and warranty costs. LPAs can also improve communication between manufacturing personnel and management, giving operators the chance to share observations and offer suggestions directly to leadership. When people see leaders taking time and interest in the work they do, they're more likely to speak up and engage with upper management.

Audits can be customized to suit operational needs. They can be short and quick, with yes or no answers, that take about 10–15 minutes to complete. Simple versions of LPAs can make it easy for people to perform one, especially those who are less familiar with the process or work in a support function. The audits should be continuously updated to reflect recent quality issues, system failures, and critical process steps that need constant monitoring to ensure compliance. Individual auditors can be selected from different departments such as operations, human resources, and finance. The diversity of auditors and frequency at which they perform audits should align with their level and scope of knowledge and responsibility. This can be defined by their standard work routines.

Compared to traditional inspections and audits, LPAs provide an additional layer of control to catch significant process variations before impacting product quality, delivery, or employee safety. Audits can help reinforce desired behaviors with different types of audits taking place multiple times daily or weekly to stress and reinforce the importance of manufacturing system compliance. Beyond influencing behavior, these audits can be used to hold people accountable from the "hard data" collected to identify who is following their standard work instructions and routines and who may need one or two nudges to do so. In cases where exceptional performance is evident, special recognition can be bestowed upon employees.

When implementing LPAs, consider some of the pitfalls to deployment including insufficient resources to plan and execute the audits, low audit completion rates due to a high demand to conduct audits, data integrity issues because people don't treat LPAs seriously, and automatically checking "yes" for an audit question without proper understanding or clear evidence. Outdated checklists, resulting from changing risks and undocumented lessons learned, can also degrade audit integrity.

If you are implementing an LPA system for the first time, consider automation to facilitate execution. Benefits to automated systems include:

■ Reduced administrative overhead provided by automated email reminders and direct links to electronic checklists.
■ Improved resource efficiency with less manual tasks.
■ Instant visibility to data. Mobile smart phones and company-issued tablets make audit data immediately available for review and response.
■ Increased audit completion rates. Real-time dashboards let organizations communicate audit results and observe when audits are not being completed.
■ Data and information management.
■ Easier document change control.

Finally, it's important to ensure every LPA is part of a closed-loop process by addressing minor non-conformances on the spot, capturing larger issues on a formal action item list for a follow-up, and adding new audit questions based on best practices and lessons learned. Reference the Process Audit Sidebar for questions to assess process and manufacturing system integrity.

SIDEBAR: STANDARD PROCESS AUDIT QUESTIONS

Process audits can be performed on nearly any process with a documented procedure or standard from which compliance and effectiveness can be assessed. Sometimes it's worth asking basic questions to determine if a standard exists, if it's appropriate, if employees are aware of it, if it is understood, if it is being followed, and if it is being confirmed as effective by the results. The following questions can be used to assess distinct processes and system robustness for compliance and effectiveness:

■ Does a standard exist and is it available?
  ■ Can you see or touch it?

- Is the standard appropriate for the work being done?
  - If you followed the standard, will it achieve intended objectives?
- Are employees aware of the standard?
  - Can they tell or show you where it is?
- Do employees understand the standard?
  - Can they explain the what, how, and why of the standard?
- Do employees follow the standard?
  - Check for deviations from standards.
- Is the standard confirmed?
  - Are periodic audits performed to confirm the standard is continuously followed?

Performance Reviews

Performance reviews are periodic assessments of operating performance that provide an opportunity for management to evaluate the "health" of process operations by monitoring KPIs and their historic trends. At a minimum, the team is looking for process stability across all critical process activities. Deviations to expected performance or targets should be highlighted, understood and actions taken to address degrading performance when detected. Generally, positive KPI trends indicate a robust system that is continuously improving.

When preparing performance data, consider standardizing the format for data presentation across the facility. Standardized data helps to facilitate quick and easy interpretation of information for understanding and clear action. It's recommended that performance reviews occur at regular frequencies, allowing people to prepare for the review to ensure efficient and effective meetings. Typical topics covered during a Performance Review may include:

- Production output trends.
- Process yield.
- Process audit findings.
- Top (5) quality issues and actions.
- Overall Equipment Effectiveness.
- Customer returns/complaints
- Problem-solving status.
- Strategic project and process improvement activities.

The primary objective of performance reviews is to provide management with the data and information they need to effectively manage the manufacturing system and realize the strategic objectives for operations. This information is also used to make decisions, prioritize problem-solving, and direct operational activities.

We discussed the concept of data analysis in the previous chapter. What's important to remember during shop floor monitoring is that the very act of monitoring involves data and information analysis. Blindly looking at visual controls without processing the data is meaningless. Data, by its very nature, requires analysis before a decision is made or an action is justified. In certain circumstances, more or less data may be required, in which case, the frequency of monitoring may need to be adjusted accordingly.

In summary, process monitoring is an on-going activity needed to understand current conditions and ensure the process is being implemented as intended and performing to expectations. It's a mechanism to ensure policies and procedures are followed, process deviations are being identified, and system abnormalities are being highlighted, evaluated, and effectively managed. We end this chapter with key points. Additional tips for shop floor monitoring can be found in Appendix 4.

Key Points

- Monitor KPIs for trends and abnormalities.
- Monitoring is an on-going activity that looks for deviations from planned activities, operating standards, and performance targets.
- Continuous monitoring provides insight into system robustness and operational efficiency.
- Spend time each day on the shop floor to heighten awareness of on-going activities and interface with employees in a coaching, problem-solving, and decision-making capacity.
- Monitor for deviations from standards so problems can be identified and corrected immediately.
- Strong leaders spend time on the shop floor observing the process and coaching employees.
- Monitor bottlenecks and problem areas closely to maintain process stability and continuous flow.
- Gemba walks are used to create awareness, enhance communication, develop people, and monitor the process for deviations.
- Gemba walks provide a real-time opportunity to take countermeasures in response to deviations, avoiding losses.
- Shop floor monitoring can be complemented by electronic displays of performance data.
- Monitoring often requires manual or electronic record keeping for evidence that certain activities were completed on time and as documented.
- Process monitoring serves as a mechanism for assessing system performance to operating standards and verifying that system outputs are meeting stakeholder expectations.
- An inspection is an examination of a work product to determine if it conforms to standards. Inspections may be called reviews, audits, and walk-throughs, among other terms. Inspections and audits are conducted to verify work compliance with operating requirements.
- Performance data and reports are evaluated against agreed to requirements, providing management with information about how effectively operational activities are achieving process objectives.
- The more eyes observing, the more improvement opportunities you are likely to find. Everyone sees things differently and different things!

3.3

Manufacturing System Control

If you always do what you've always done, you'll always get what you've always got.

~ Henry Ford

Objective: Achieve and maintain manufacturing system stability, capability, and control.

Overview

A fundamental objective of shop floor management is to achieve process stability through defect elimination and process capability by methodically reducing sources of variation until process output consistently meets operating targets. Process stability is realized when process performance becomes predictable over time and process capability is evident when process output consistently meets engineering or customer requirements. Process control, realized through daily standard work routines (SWRs), is then employed to maintain process stability and capability while establishing the baseline for sustainable process efficiency improvements.

An unstable process often results from management indifference or neglect of the manufacturing system. Chaos reigns in the absence of a structured approach to system management as people work to do their best with the knowledge, support, and resources available. Chaos can be avoided when you have key individuals who "know what to do" and work to maintain operations at a sufficient level to satisfy stakeholder expectations. However, as those people leave their positions or the company, process stability can be quickly lost. This is why a manufacturing operations manual that clearly defines operating practices and procedures helps bring structure and focus to the activities required for maintaining operational stability, capability, and control.

Continuous improvement is also considered as an essential component of modern-day shop floor management. When there is little desire to challenge the status quo, change becomes difficult. If the status quo is no longer tolerable, due to rapidly changing industry dynamics, it's likely time to change. If you are serious about improving shop floor performance, you must first understand what's "ailing" the existing manufacturing system. To do so, consider the following questions:

- Does a clearly defined system exist for managing the shop floor (e.g. who does what, when, where, how, and how often)?
- Is the existing shop floor system capable of meeting operational expectations and deliverables?
- Are key employees aware of the shop floor system; do they understand it and know how to properly execute it?
- Do those responsible for the system execute, monitor, and control it, as defined? If so, are they getting desired results?
- Is system functionality and effectiveness periodically confirmed?

Your response to these questions will likely determine your next steps. Remember, change requires knowledge of where you are relative to where you want to be. If you decide it's time to change, consider what is the desired state of operations. What's currently not meeting expectations? What needs to improve? What's the plan to change so that expectations can be realized?

In the following chapter, we will explore the activities and work routines required to maintain operational control in today's manufacturing environment. The items discussed will focus on the fundamental practices of daily manufacturing management. Although these practices are universal, every operating system needs to be customized to suit the working habits of employees and the company culture.

Manufacturing System Control

Proper control of the manufacturing system relies on effective procedural execution and monitoring of visual controls, to identify deviations from standards and trigger

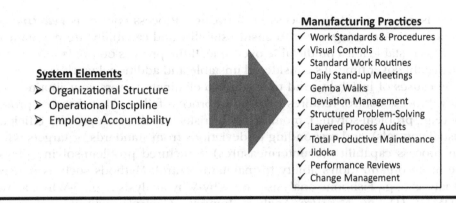

Figure 3.3.1 Manufacturing Practices and System Elements.

actions to maintain consistent and predictable output performance. Good manufacturing system control not only focuses on the process to produce products but on the operating system created to support the people, materials, methods, equipment, and environment required for consistent product assembly, testing, packaging, and delivery. The operating system must be in sync with the assembly process to ensure quality products are delivered on time and in the quantities requested by customers. We will consider process control, from the perspective of maintaining product line stability and capability, as well as the manufacturing system, within the context of daily shop floor management. Let's start with a focus on maintaining process control of product lines and follow up with a broader view of activities to control the operating systems to ensure system integrity and robustness.

Process control requires a conscious effort to identify and execute the daily, weekly, and monthly activities and routines necessary to maintain a consistent and sound process. Standard practices and organizational discipline are necessary to maintain and sustain a well-running operation. This is the starting point for lean manufacturing. However, before we can apply lean concepts for process improvement and optimization, we must first ensure the fundamentals of process control exist, are practiced, and systems are in place to sustain them. Unfortunately, process control is one of the most important yet, neglected activities in manufacturing today. It's boring, tedious, and time consuming but vital to the long-term health and integrity of operations. In the context of daily shop floor management, some of the basic practices and system elements needed for manufacturing system stability and sustainability can be found in Figure 3.3.1.

Process Control

Process control is the act of monitoring and controlling a process for stability and capability. A process is **stable** when its output is consistent and predictable over time and **capable** when the process output consistently meets customer requirements. ***Process capability can't exist without process stability***. Process control is maintained by performing daily and weekly work routines intended to monitor the process for anomalies and deviations to standard practices and implementing countermeasures to maintain process performance. Process control is *reactive* and requires *deviation management* to ensure stability and capability are maintained.

A process is stable if it's predictable over time. If the process output is inconsistent or unpredictable over time, it's considered unstable and additional work is needed to assess the causes of instability and take steps to eliminate factors contributing to its instability. Process control requires reacting to process failures that threaten process stability (e.g., problem-solving), monitoring for risks that threaten process efficiency (e.g., visual controls), and responding to deviations from standards or targets which threaten process capability (countermeasures). Structured problem-solving plays a key role in an organization's ability to maintain control. Methods such as Cause & Effect analysis (e.g., Fishbone analysis) and Why-Why analysis (e.g., 5-Why), as well as the DMAIC (Define, Measure, Analyze, Improve, Control) methodology, can be used to reveal the root causes of process destabilizers which can then be targeted for elimination.

Once stability is achieved, output results must not only be consistent but also consistently meet customer requirements to ensure continued satisfaction. If process requirements can be achieved reliably, this helps ensure the same for customer requirements. If excessive process fluctuation exists relative to specification limits, steps must be taken to reduce process variation to within acceptable limits necessary to produce suitable product. If process outputs can't consistently satisfy customer expectations, additional work will be required to reduce process variation within acceptable boundaries of process capability.

In review, process stability is reflected in predictable product output demonstrated over time, while capability is the ability of systems to deliver results that consistently meet stakeholder expectations. Control is the result of maintaining long-term stability and capability. It can be achieved through the execution of specific and periodic work routines and management of deviations from standards and targets. Executing daily, weekly, and monthly routines help ensure that manufacturing systems continue to perform as intended. This is done by assigning daily work routines to everyone hired to maintain process control and driving continuous improvements.

Employee accountability to these routines is assured by periodic verification that the routines are being implemented at the frequency specified to ensure proper and effective execution of the manufacturing system and confirming results reflect intended output performance. Process control relies on competent people, reliable equipment, material availability, capable procedures, and a stable work environment. See Figure 3.3.2 for a process control and improvement view of shop floor management.

Manufacturing System Integrity

Control extends beyond the product assembly processes. It must touch all activities needed to maintain the manufacturing system. This includes practices such as facility up-keep, Human Resource administration, material handling and equipment maintenance. A series of reactive and proactive activities are required to ensure system predictability and integrity over time. Practices such as maintaining workplace organization, following operating standards, executing SWRs, monitoring visual controls,

Figure 3.3.2 A view of shop floor management.

performing layered process audits, and exercising deviation management comprise the foundation of the manufacturing system and must become part of daily shop floor management activities. Periodic verification of system effectiveness also needs to be an integral part of shop floor routines. Output performance and key performance indicator (KPI) trends are good indicators of system integrity which can also be observed through employee engagement during daily meetings, process audits, and Gemba walks.

Daily monitoring and control of key processes is the job of every individual working in manufacturing. If significant deviations are identified, appropriate actions must be taken to understand their root cause and implement a solution to bring the system back to baseline performance. This is deviation management which is often complemented by problem-solving as a primary mechanism for maintaining control.

It's important to highlight, stop, and fix problems quickly. When problems occur, they cause disruptions and can degrade process performance, potentially delaying commitments made to customers. Effective problem-solving may require observation and analysis, sometimes slowing down the process until the cause is identified and a countermeasure taken. It's important to get quality right the first time by building quality into the system and using simple visual controls to highlight concerns when initial quality can't be assured. Looking at problems as opportunities to learn about system and process performance can lead to creative solutions and continuous improvement. In short, system management and control requires the engagement of people to observe the processes within which they work and react to deviations that are likely to disrupt performance results. Manufacturing system integrity is a choice made by people to make it a reality.

Standard Work Routines

SWRs, as previously discussed, play a key role in daily shop floor management by defining the current scope of activities necessary to execute, monitor, and maintain manufacturing operations. This includes products to be manufactured, processes to be followed, resources needed to do the work, equipment maintenance, and the discipline to complete scheduled work, on time and correctly. While operators must follow product assembly instructions, every other employee should have a set of SWRs, they are expected to follow, as part of their value contribution to managing the shop floor. SWRs define the periodic, repetitive tasks employees are expected to perform to ensure a stable and predictable process over time. Every employee should have a clear understanding of what they are expected to do, when they are expected to do it, and why it's important to do it.

For the most part, shop floor activities are not glamorous but a clearly defined and properly executed set of routines, proactively executed, helps to maintain the status quo, until a better way of doing work is proven effective. Consider SWRs as the everyday roadmap for running the shop floor. They are one of the building blocks for process optimization and an essential element of operational excellence. Chapter 3.1 provides more detail on the topic of SWRs.

Deviation Management

Deviation management is a fundamental and effective shop floor management methodology. It's a systematic instrument of leadership to maintain process control and product quality. It's the result of organizational structure, operational discipline, and employee accountability and can only happen effectively when standards, visual controls, and SWRs are in place to facilitate it.

Deviation management requires monitoring the process for variations from standards such as machine setup parameters, material specifications, output targets, and work instructions. Significant deviations can cause process instability which may lead to defects, impacting product quality and result in customer complaints. Thus, corrective actions can address a deviation and steer the process back into a stable state.

Deviation management happens when procedures are executed, systems are monitored for nonconformities, and significant abnormalities are corrected through countermeasures. These activities are facilitated through the deployment of operating standards, visual controls, and SWRs including 5S, layered process audits, Gemba walks, and performance reviews. Process control requires the discipline of systematic problem-solving to help identify and eliminate the root causes of deviations and track implemented solutions for effectiveness and reoccurrence.

Deviation management starts with the visualization of process standards and targets. Visuals can be pictures, charts, and graphs that are displayed for easy viewing. These can include a board that displays actual production output versus target, a KPI trend chart or picture of a good versus bad part. These visuals serve as a signal or trigger during process monitoring to enable rapid identification of abnormalities and deviations from standards. 5S and visual controls can help facilitate easy and rapid detection of abnormalities. As appropriate, these concerns may be addressed through immediate action or warrant further investigation prior to identifying an effective course correction. Corrective actions identified during the deviation management process should be documented, prioritized, and actively managed until closure. See Figure 3.3.2 for the role of deviation management within shop floor management. The deviation management process can be viewed in Figure 3.3.3.

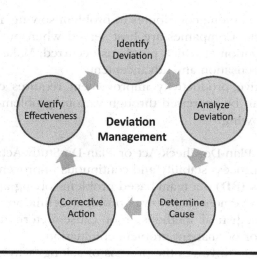

Figure 3.3.3   Deviation management methodology.

The practice of deviation management is critical to establishing and maintaining process and system controls. It must be an inherent part of daily shop floor management without which efficiency slowly degrades while chaos ensues. The following deviation management tips are intended to aid the effective execution of deviation management.

Deviation Management Tips

- Actively engage in deviation management as an integral part of daily shop floor management.
- Visually record deviations real-time to highlight issues impacting manufacturing operations.
- Look for deviations from standards during shift changes, machine start-up, and tool changes.
- Correct deviations on the spot, if possible. Assign an action item to those activities that will take more time to address.
- All actions should be assigned a priority, an owner and due date for timely follow-up.
- Identify a root cause, implement a solution or corrective action, and verify compliance to standard or procedure.
- Review the status of action items on a regular basis.
- Make timely action item closure a visual control and metric.

Structured Problem-Solving

Manufacturing people are paid to follow procedures, make improvements, and solve problems. Good problem-solving skills is one of the most valuable assets an individual can bring to a manufacturing environment. Significant operational efficiency can be realized when everyone contributes to problem-solving, not just a handful of problem-solving experts. Companies are best served when problems are solved at the level of the organization at which problems occurred. Make problem-solving an essential skill for compensation and advancement.

Efficient and effective productivity improvement requires competent problem-solving skills which can be exercised through various problem-solving methodologies including the following:

- **Deming Cycle** – Plan-Do-Check-Act or Plan-Do-Study-Act: a simple problem-solving cycle for process stability and continuous improvement.
- **Eight Disciplines (8D)** – a team-based problem-solving approach for product and process improvement. It involves problem definition, defect containment, data collection, statistical methods, root cause determination, and solution implementation for permanent problem elimination.
- **Why-Why (5-Why)** – involves the process of asking "why" multiple times until root cause is determined. It requires a logical approach and critical thinking skills to drill down to a problem's root cause and will likely require some analysis and investigative skills to do so.
- **Kepner-Tregoe Decision Analysis** – a structured approach for gathering information, prioritizing and evaluating data. It involves the four-step process of situational appraisal, problem analysis, decision analysis, and potential problem analysis with the objective of minimizing problem risk.
- **Six Sigma** – Six Sigma is a disciplined, data-driven approach for improving the quality and efficiency of an organization's operational and transactional processes. This approach uses the DMAIC methodology consisting of Define, Measure, Analyze, Improve, and Control.

Keep in mind that these are only problem-solving methodologies. It takes knowledgeable and skilled employees, able to effectively use these approaches, to determine root cause and implement permanent corrective actions that eliminate problems from reoccurring. The ability for all employees to solve problems is entrenched in an operationally excellent culture. Encourage employees to think freely, collect data, and analyze it for information. Expect them to use this information to take meaningful action. Recognize and reward individuals for exercising their problem-solving capabilities effectively.

Problem-Solving Tips

- Go and see a problem at its source, whenever possible.
- Work with reliable data and factual information.
- Put emotions aside, think logically and scientifically.
- Solve problem at the lowest possible level of the organization.
- Work in logical groups to solve problems.
- Be prepared to communicate your findings and conclusions. Anticipate questions.
- Increase organizational capability to solve problems by moving focus away from problem-solving experts and training everyone to solve problems.

- Coach others in problem-solving by solving problems real-time.
- Use a problem-solving method to ensure efficient and effective problem resolution.

Action Items

An action item list is a tool that helps individuals manage (and prioritize) daily work activities. We document and implement actions we believe will add value upon completion. Documented actions capture our thoughts and decisions. They serve as a mechanism for managing the manufacturing system, as countermeasures to deviations and help drive incremental and strategic improvements. An action item list serves to document, organize, and prioritize work tasks. It provides structure by describing the topic of interest, activity, activity owner, status, and target completion date to drive timely closure.

An actively managed list of actions is an essential tool in the shop floor management arsenal. Actions must be documented, prioritized, and implemented to ensure meaningful manufacturing control and continuous improvement. When monitoring the manufacturing system for compliance to procedures and standards, significant deviations must be addressed immediately or captured as an action to drive prompt follow-up based on existing priorities. Actions must be documented for issue awareness, communicated to relevant stakeholders, and assigned ownership so accountability can be established and maintained until completed. The assigning of due dates helps trigger timely follow-up and completion. No documentation, no follow-up, no control, no improvement.

Certain actions can be completed real-time such as observing a deviation from procedure. In these incidents, workers can be coached, on the spot, to correct an omission or misstep. Following up with the offending individual, immediately after the event, is important to ensure continued compliance and helps to reinforce the desired behavior. Certain abnormalities may take more time to understand and require an in-depth evaluation to determine root cause so effective corrective action can be identified and implemented. This is where an action item list adds value, by capturing issues for prioritization and effective management.

Generally, an action item list should be maintained for a specific department, function, or area. A more "generic" action item list can be established for the shop floor when an issue or action impacts multiple areas or aspects of operations. Whenever an action is identified, it's important to specify the context of the action. For example, was the action the result of a specific audit, poor quality, a safety issue, or an improvement opportunity. By classifying actions into groups, they can be more easily prioritized, categorized, and followed up by the appropriate owner.

Figure 3.3.4 provides an example of an action item list. Whatever format you choose, make sure it's easy to use and reference. Don't hesitate to update the format based on the needs and lessons learned of the organization. Manual recording of action items is common, but more digital systems are now available for use. Like all electronic systems, resources will be required to maintain and upgrade any system chosen for implementation.

Action Item List

Item #	Open Date	Topic	Action	Owner	Due Date	Status	Comments

Figure 3.3.4 Example of action item list format.

Decision-Making

Good decision-making is a key attribute of management and leadership. It's a critical skillset that deliberately and thoroughly considers all relevant information, facts, and options. A good decision is clearly explainable as to the logic used to make it. Decision-making, in complex situations, requires understanding the significant factors that affect decision results. This includes evaluating the decision options, establishing priorities, anticipating outcomes, and recognizing the potential risks and uncertainty that may stem from a decision. Key traits of a good decision-maker include being open minded, listening to all relevant viewpoints, knowing expectations, considering all options, critical thinking, and learning from past experiences. As the adage goes, garbage in, garbage out.

Constant decisions are being made on the shop floor. The quality of those decisions is best when reliable data and information are used to make the decisions. The best decisions are made when the decision-maker is fully aware of the situation, considers all options available and reflects on the potential outcomes of the path chosen. A decision can result in several different outcomes, not all of which may be obvious. Start by considering the pros and cons of each decision. If helpful, talk it out. Seek different perspectives. Consider alternatives. Ask for more data and information.

Manufacturing decisions, relative to technology, need to consider the cost of hardware, software, and the expertise required for installation and maintenance. The benefits of implementing certain technologies must outweigh the cost and potential on-going effort to support and maintain it. Many shop floors operations are fast-paced, lean manufacturing environments with minimal resources to pursue technological advances. In these cases, a thorough evaluation of a technology's impact on operations needs to occur and used to justify a decision. A final decision should be evaluated for its impact to the bottom line. If a decision's outcome does not positively affect safety, quality, productivity, or cost, think twice before moving forward. If pursuit of a technological solution can't be justified, simply wait to see how the technology evolves over time and reevaluate the topic at a later date.

It's important to make decisions that benefit the organization. A cost savings decision in one area or department may negatively impact another. When making decisions, consider its effect on the entire manufacturing system. Decisions made in isolation can undermined the ability to make good, cost-effective, and beneficial decisions. However, once a decision is made and executed, monitor its impact. If the outcome is not as expected, review the inputs used to make the decision, lessons learned and consider collecting more data. Don't stick with a bad decision, if the outcome does not satisfy expectations. Review your alternatives.

Poka-Yoke (Error Proofing)

Poka-yoke is a Japanese term for mistake proofing or error proofing. It's the application of a mechanism, implemented in manufacturing, to help workers avoid making mistakes that can lead to defects. The objective is to detect, prevent, correct, and draw attention to human errors as they happen, eliminating defects at the source. Poka-yokes are everywhere such as in the kitchen microwave that automatically shuts off when the door is opened or an elevator door that automatically retracks when a passenger enters during door closure. Look for opportunities to implement poka-yokes in manufacturing, especially in processes where errors are prone to occur. For example, if a component can be placed incorrectly in a fixture, look for ways, such as using a notch, to ensure proper orientation.

Poka-yokes can be applied in several different ways. One type of approach is to test the product's shape, size, or other physical attributes for a known condition which, if not met, will trigger a warning or stop the process. Another type of poka-yoke will alert the operator if a certain number or sequence of movements are not completed. Regardless of the method applied, prevention over detection is generally preferred. Detection still requires operator intervention to prevent a potential defect from occurring.

Defect avoidance and prevention reduces the overall costs of mistakes. Other benefits of adopting a poka-yoke mentality can include a reduction in training hours needed to prevent defects, reducing resources needed for quality control, decreasing an operator's work burden, and avoiding defect escapes to the customer.

Jidoka

Jidoka is a lean methodology based on the principle that a machine or process is stopped automatically and alerts the operator when an abnormality or error is detected. The problem is expected to be corrected immediately or contained until a root cause corrective action can be identified and implemented to prevent reoccurrence. The objective is to stop defects at the source to prevent the scrapping and rework of units downstream of the offending process. Jidoka is an "intelligent" form of manufacturing, within the scope of deviation management, which requires human intervention to correct a problem and restart the process.

A trigger to stop the process is usually based on the type and number of defects that occur within a given time period (Jidoka limit). Since the priority is usually to restart the process as soon as possible, a containment or corrective action is typically implemented. A containment would be needed if the problem could not be eliminated immediately. In this case, the root cause must be investigated, identified, eliminated, and the containment removed before the process can resume as normal.

Poka-yokes can be deployed as part of the Jidoka strategy to prevent defects. If defect prevention is not possible, the mindset should be to stop the process immediately, upon defect detection, to ensure quality at the source and prevent the build-up of defects that may lead to scrap, rework or a manufacturing escape to the customer. Manufacturing excellence is reflected in the ability to stop and correct defects when they occur.

Jidoka limits may be set high for a new process since limited time and resources may prevent a quick response to every defect generated. However, with the passage of time, defects will be eliminated, reducing their frequency of occurrence. As a result, Jidoka limits should be subjected to periodic adjustments, since the amount and type of defects will decrease through problem elimination. The goal is a "Jidoka limit of one", where the production team immediately reacts to every defect that occurs. As more manufacturing errors are detected and eliminated, process stability will improve, allowing shop floor personnel to "supervise" multiple machines or

Figure 3.3.5   Jidoka process flow.

areas for quick response to issues, at a reduced labor cost. If properly implemented, Jidoka will have a positive impact on overall equipment efficiency (OEE). See the Jidoka process flow in Figure 3.3.5.

Change Management

Change management is a systematic approach to transforming an organization. The purpose of change management is to identify the need for change, validate change, implement change, and control change, in addition to helping people accept and adapt to change. It typically involves updating processes, procedures and managing people so they can effectively integrate change into a new or revise way of working. Managing change puts an individual or team in a position to influence its implementation and outcomes.

The change process starts by understanding the kind of change being requested so it can be effectively proposed, and if accepted, deployed. An important part of the change proposal and deployment process is to communicate why a change is necessary and what will be achieved upon successful implementation. The change management process also involves building consensus and commitment to delivering change that results in real and sustainable improvement. All improvements involve change but not all changes bring improvement.

A process or system can only be improved if the standard to which it's being held is stable and measurable. Ideas for improvement can come from many sources including employee suggestions, observations, experimentation, and lessons learned. All change proposals should be considered but prioritizing accepted changes for implementation is key when there is limited time and resources to manage deployment. Prioritization will ensure the most meaningful and impactful changes are implemented soon after approval, to maximize their benefit to operations.

Piloting a change before implementation is recommended to evaluate a change's impact on manufacturing and to verify its effectiveness in meeting expectations. This is especially true if reversing the change will be difficult, costly, or time consuming. Deployment of approved changes should follow a structured approach to optimize deployment efficiency. A simple method such as the plan-do-check-act cycle may be adequate in most cases. A project management approach may be needed when a more substantial change is being deployed. Upon implementation, all changes should be validated to confirm they are meeting the intent for which they were approved.

When looking to implement a change, talk to key stakeholders, especially those that will be significantly impacted by the change to obtain their feedback and support. Successful change is not only about documentation and training but often more about acceptance and commitment. If people don't believe in a change or see how it can benefit them, they are unlikely to support it, allowing the change to die a quiet death. Communicating the need and benefits of change is key. Lack of awareness is another reason why change fails. When a change has been approved, inform all key stakeholders. Once deployed, confirm realized changes remain in place and effective over time. Ensure unapproved changes do not get implemented. The following basic steps can be used for change management:

- Identify change opportunities.
- Propose a change.
- Assess change's impact on existing processes/system.
- Demonstrate/prove out effectiveness.
- Accept/reject change.
- Prioritize approved changes for implementation.
- Implement change.
- Follow up (verify sustainability).

Production Change Points

Production change points are periodic events that occur in production when software, equipment, tooling, and procedures are changed to accommodate a new production model or variant based on the production schedule or customer demand. These are planned changes that happen during the life cycle of a production line. Production change points commonly occur at workstations when changes are needed for model setup and routine changeovers. Tooling changes, for example, often occur during changeovers but can also happen when control limits, used to monitor tool wear, trigger a tool change.

Other examples of change points include changing a bowl feeder to a new part number or realigning a detection sensor for a new part configuration. Unfortunately, these changes or change points create manufacturing risk that can cause quality issues if not properly controlled. To mitigate risks, production changeovers should be properly controlled by following standard work instructions and monitoring output quality for abnormalities during production restart.

Bottleneck Management

A bottleneck in manufacturing occurs when the workflow to a production process exceeds its capacity to handle it. The bottleneck sets the rate of production by being the slowest workstation in the process. Bottlenecks can limit the capacity of production output. If the bottleneck rate disrupts the production takt time, preventing the system from achieving its goal, it becomes a constraint. System constraints will cause production delays and can increase manufacturing costs.

One approach to bottleneck management is to release work into the system at a rate consistent with the bottleneck's capacity. This helps synchronize production to the constraint while minimizing inventory and work in process (WIP). Another option is to place a buffer before the constraint to keep the bottleneck station busy. This helps to maintain maximum output performance since any lost time at the bottleneck results in a further capacity drop.

WIP will start to increase if other workstations start producing at a rate greater than the bottleneck workstation. If you want to increase the capacity of a system, focus on increasing the capacity of the bottleneck. Consider offloading work to another station or increase resources at the constraint to increase system capacity. As a principle, systems will always have bottlenecks. How you manage those bottlenecks will dictate your operating capacity and efficiency.

Additional Manufacturing System Controls

There are other process control methods, tools, and techniques that have been used extensively in various industries. Although they may be excessive for some industries, they have proven to be very effective for those that require a high degree of structure and discipline in their operations to produce reliable products. This is not an exhaustive list but supplements what has already been presented.

- **Process Flow Charts** – a visual map that outlines a series of sequential activities that turns raw material into finished goods for delivery. Flow charts can help with process monitoring and control.
- **Process Control Plans** – a document that defines key process input parameters and information used to minimize process variation and deliver products that meet customer requirements.
- **Process Failure Mode and Effects Analysis (PFMEA)** – an analytical tool used to identify and evaluate potential failure modes and their corresponding effects on process operations. It's used to assess the likelihood and impact of undesirable events on process performance and prioritize high-risk items for mitigation.
- **Control Charts** – a graph used to visually monitor process changes (variation) over time and trigger action when unacceptable process behavior is detected.
- **Parameter Cards** – documents the setup and changeover parameters for a production line. The card helps maintain process control by serving to verify the correct parameters are set prior to the start or restart of production.
- **Checklists** – a list of items to be completed. It helps ensure certain process activities are considered or done. For example, it can outline the steps required to start up or shut down a piece of equipment.
- **Schedule** – defines a plan of activities and their corresponding time for completion.
- **Logs** – a record of events often used to confirm an activity or task was completed.

In summary, system control is not just a good idea, it's an essential practice of manufacturing excellence that needs to be defined, exercised by all, and continuously improved based on experience and lessons learned. The activities involved in control take on increasing importance as products and their corresponding processes become more complex. A visual of manufacturing system control methods is presented in Figure 3.3.6.

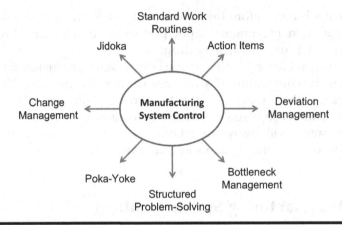

Figure 3.3.6 Manufacturing system control methods.

As usual, I will leave you with some key points to consider during your leisure time.

Key Points

- Leaders must allocate the time and provide the resources to ensure system controls are defined and properly executed.
- A deviation can be anything that does not meet a standard or expectations.
- Deviation management compares actual process performance to a standard or target and works to bring unacceptable process deviations within acceptable limits or control.
- SWRs involves monitoring and controlling key aspects of the manufacturing process.
- Managing problems aggressively will build confidence, increase personal satisfaction, and provide an enhanced sense of control over one's work!
- Monitor bottlenecks and problem areas closely for abnormalities and deviations that can destabilize a process or system.
- Control looks to identify ways to eliminate causes of *unsatisfactory* results.
- Process control starts with a clear understanding of the cause-and-effect relationship between process inputs and activities relative to output performance.
- Documenting problems that don't lead to corrective action is waste.
- Ensure all employees are working to control and improve their area of responsibility.
- Exercising control requires that standards are in place to create a baseline from which control can be confirmed.
- Deviation management is required for process control as change management is required for continuous improvements.
- Good decisions are achieved through common sense and good judgment.
- Although we can't control the changing world around us, we can control how we respond to it.

- The act of process monitoring and control ensures the process remains in compliance with operating standards.
- Quality at the source requires operators to stop the process when a defect is detected, and act immediately to prevent additional waste.
- The purpose of implementing Jidoka is to detect and diagnose a product defect or process malfunction and correct it immediately.
- Systems that are not followed as expected, fail to deliver on expectations.
- Minimize resistance to organizational change by involving key stakeholders in the change process.
- Process control is the foundation for sustainable efficiency improvements, optimization, and manufacturing excellence.
- Taking ownership for needed improvements leads to stable systems and improving performance trends.

3.4

Manpower

Great things in business are never done by one person, they're done by a team of people.

~Steve Jobs

Objective: Treat others as you would want to be treated.

Overview

Manufacturing needs good people. You can't run a manufacturing facility without them. People are hired to maintain the facility, purchase and handle materials, assemble products, run equipment, solve problems, satisfy customers, work with suppliers, and execute procedures. Although robots are prevalent in some of today's shop floor environments, you still need people to install, program, and maintain them. People remain the brains behind manufacturing operations. However, if you look at manufacturing from a different perspective, labor is one of the key expenses, besides material, that determine product costs. This makes minimizing labor costs a critical

DOI: 10.4324/b23307-17

focus of many modern-day operations, in high-cost countries. To remain efficient and competitive, optimizing labor utilization is essential. Labor utilization and productivity are affected by multiple factors including some of the following.

People Availability

It goes without saying, no people, no production. Hiring people can be challenging when unemployment is low or, in more recent times, when people are afraid to work due to the fear of becoming ill during a pandemic. At times, it may be necessary to raise wages or ensure proper safeguards are in place to minimize exposure to the spread of a contagion.

Consider what benefits will attract people to your company versus others and exploit those benefits when promoting available opportunities. Will higher wages, an educational advancement program, a bilingual work environment or subsidized day care attract more people to your company? Sometimes "thinking outside the box" is simply a matter of understanding what people want and need when choosing a job or becoming a long-term employee.

People Flexibility

When material supply is tight or material transport is delayed, especially in a just-in-time work environment, a lack of material will disrupt the production planning schedule. This may require the movement of available people to other production lines or possibly sending people home due to a lack of material need to build product. In these circumstances, cross-trained employees can increase operational flexibility.

If no work is available to reallocate resource, before sending people home, consider different ways to utilize their available time. Cleaning, training, equipment maintenance, and other "spur of the moment" activities should be considered. These options work best when the manufacturing team has contingency plans in place for such events. When people's livelihoods are at stake, eliminating material supply issues, among other operational disruptors, is extremely important. People start leaving companies who can't guarantee them a steady paycheck when needed to live.

Workload Balancing

A critical element of continuous process flow is workload balance, facilitated by standardized work tasks. The goal is to create equal time intervals among the workstations within a production cell. A balanced workload, among employees, helps to maintain a steady rhythm, contributing to a continuous workflow and the creation of an equitable work environment. It's important for people to see others sharing in the workload when equal pay for equal work is expected.

Manufacturing Gauges and Tools

Tools are intended to make work faster and easier for people to complete. The right tool, at the right time, can improve operational efficiency. The right tool, in the right hands, can also support product quality and timely delivery. Gauges and tools need to be available, in good working order, and properly maintained to ensure maximum benefit.

Spending time looking for a tool, fixture or gauge, waiting on its calibration or repair is considered waste and will decrease labor efficiency while increasing operating costs. Take the time to monitor and control essential production equipment such as gauges, fixtures, and tools. Often, the benefits of doing so far outweigh the consequences of not doing anything.

Employee Motivation

Contented employees tend to be motivated employees. When people have the resources, direction, and support they need to do their work, they are more likely to do good work. Pride in one's work is often reflected in their actions, interactions, and results. Equal pay for equal work also sets the stage for fairness, trust, and an engaged workforce. When walking the shop floor, gauge people's emotions and look for opportunities to boost employee morale.

Training and Experience

Training elevates knowledge, understanding and employee capabilities to do the work for which they are tasked to do. Experience reduces missteps, builds confidence, and improves operational efficiency. Training is an investment in people that pays dividends in product quality and customer satisfaction. Remember, training and experience comes in many forms including employee coaching and mentoring, engaging in workshops, on-the-job training, taking classes, observing others, attending academic courses, and through self-study. Don't ignore the need for time and money to train and develop people when preparing strategic plans and developing a budget for shop floor operations. Although training costs money and time, untrained people will likely consume money and time through operational inefficiency and defect generation.

Workplace Design and Ergonomics

Workplaces should be designed with people in mind. This may sound obvious but often does not happen. It's not difficult to identify opportunities for improvement when observing an operator in motion as they execute their work instructions. Simple changes to a workstation like improved lighting, tilting a box to pick up parts easily, or placing a fixture in a convenient spot for access during the assembly process can add up and make an operator's job easier and less taxing to perform. Small, simple improvements, in light of performing repetitive activities over an 8-hour shift, can have a significant impact on operator health and wellness. Always look for opportunities to improve workplace ergonomics especially for those engaged in repetitive work.

Working Conditions (Environment)

Safe, healthy, and comfortable working conditions are a minimum expectation of any considerate shop floor operation. When these conditions don't meet a minimum standard of expectation, employees can become quickly dissatisfied. Work place design, cleanliness, orderliness, and ergonomics can go a long way to maximizing employee satisfaction, retention, and efficiency. Working conditions related to the following topics should also be considered in the spirit of continuous improvement:

- Work hours.
- Exposure to environmental hazards.
- Work responsibilities.
- Work-life balance.
- Workspace.
- Heavy lifting (physical strains).
- Job aids.
- Legal rights.
- Compensation.
- Profit sharing.
- Job security (stable employment can be reinforced through regular employee communication and feedback).
- Benefits (insurance, parental leave, daycare, wellness programs, Paid Time Off, sick leave, education support, etc).

Consider company policies, procedures, and practices and how they are applied to manufacturing personnel. Look to improve labor efficiency by observing how employees are interacting with their environment and each other. Talk with workers to gauge their emotional state and use lessons learned from these activities to improve the shop floor environment.

Autonomous Maintenance

Autonomous maintenance, one of the features of Total Productive Maintenance (TPM), is a cost-effective approach to enhancing operational efficiency. Qualified operators are assigned the responsibility of maintaining their own machines within defined parameters. Since operators have a good understanding of how their equipment works and behaves, they are more likely to notice a change in equipment operation or performance. Employees willing to take on this responsibility are trained in understanding their equipment, allowing them to perform regular maintenance, manage changeovers, and sometimes complete minor repairs, if within their scope of training.

Assigning this responsibility to equipment operators can significantly reduce the downtime associated with waiting for a technician to arrive and repair a machine. This skillset, complemented with a strong productive maintenance program, can radically increase equipment availability and overall equipment effectiveness (OEE). Autonomous maintenance can also provide a career path for operators that look for more responsibility, a pay increase, and exposure to better jobs opportunities. This also frees up the maintenance group to focus on more complex problems and project work for long-term equipment upgrades.

Labor Efficiency

Manufacturing labor efficiency reflects how capable the workforce is in completing a task relative to a standard. For example, if a factory production line standard is to produce 60 units in 1 hour with three people, this would be considered 100% labor efficiency if this standard is met. If 63 units were produced in 1 hour using three people, you are more efficient. However, if you only produced 54 units with three people within an hour, you would be considered less efficient. Labor efficiency is calculated by dividing the standard labor time to compete a task by the actual time worked to complete the task, multiplied by 100 or

$$(\text{Standard Labor Time} / \text{Amount of Time Worked}) \times 100 = \text{Efficiency}$$

Thus, if the standard labor hours to complete 300 units is 42 hours and it took 51 hours to complete, efficiency is (42/51) x 100 = 82%.

Efficiency can be influenced by company culture, work standards, ergonomics, workplace design, employee experience, and a host of other factors. Opportunities to improve labor efficiency may be realized through equipment upgrades or the application of new technologies such as cobots, automated guided vehicles (AGVs), or robotic process automation. These topics will be discussed in Part 5.

Employee Safety

According to the national safety council [1], the top five injuries to manufacturing workers are due to: contact with an object (40%), overexertion (24%), slips, trips and falls (19%), repetitive motion (8%), and contact from harmful substances (6%). Manufacturers with the best safety records prioritize safety. This culture needs to be driven and demonstrated by leadership, reinforced by management, and practiced by frontline workers with the mindset that they are responsible for their own safety.

Safety should be engineered into equipment and processes to reduce human error, accidents, and injury. Working conditions should take into account ergonomics, scheduled breaks, and the impact of repetitive work causing fatigue, injury, and sometimes illness. Small, intentional production breaks can help mitigate extreme fatigue, especially if combined with fatigue-relieving exercises. One way to help improve safety is to create checklists and initiate safe practice programs. Occupational Safety and Health Administration (OSHA) regulations are a good starting point for elevating safety consciousness. However, a detailed checklist of potential fire, electrical and ergonomic hazards can be highlighted and communicated to heighten awareness and serve as a basis for periodic audits. The use of technology such as the internet of things (IoT), automation and data analytics can also be leveraged to improve safety. Automation and robotic devices are being employed to do hazardous work, once performed by humans, while IoT sensors can offer better information on machine operation and process performance.

In this modern age of manufacturing, sensors and artificial intelligence can now be paired and, through the application of algorithms, deployed to detect when workers may be more susceptible to heat stroke, chemical exposure, or other unfavorable working environments. Sensors can be installed to track both human and machine working conditions, streaming data that gives manufacturers' an opportunity to make quick decisions and facilitate on-going safety improvements to reduce operator and operating risks.

SIDEBAR: COVID-19 IN THE WORKPLACE

During the pandemic, the company I worked for took exceptional measures to follow the Centers for Disease Control and Prevention (CDC) guidelines to protect employees from getting or spreading the Covid-19 virus. They put their best efforts into making workers comfortable coming to work and working with others. Masks were mandated and provided to everyone, distances were respected wherever possible, and when not possible, transparent barriers were erected to minimize exposure of employees to each other. Temperature monitors were also installed at the front entrance to check employees for abnormal conditions. For some, the workplace became a place of refuge, outside of which exposure to the virus was less certain.

Many employees were comforted by the fact that they were less likely to bring the virus home from work compared to other family members who may have been required to work or interact in less controlled environments or in professions with increased exposure potential. Although there was no guarantee against catching the virus, the company took all standard and practical precautions to contain and prevent the spread of the virus to others. Protecting people in this way makes good business sense.

Key Points

- Minimize employee absenteeism by making your company an employer of choice.
- People know their work activities the best and are well suited to understand or assist with problem-solving in their area of responsibility.

- Strive to utilize about 85% of an employee's available time, allowing some time to address unexpected activities and events. Standard work routines can be used to defined and prioritize shop floor activities within the scope of an employee's work time.
- Strategically plan and budget to train and develop people.
- People are more flexible than robots, but robots are more consistent and repeatable than humans.
- Design for people's comfort and ease of work.
- Strive for a balanced and fair workplace.

Source

[1] https://injuryfacts.nsc.org/work/industry-incidence-rates/industry-profiles/

3.5

Material

The more inventory a company has, the less likely they will have what they need.

~ Taiichi Ohno

Objective: Maintain a continuous flow of good material to the production line.

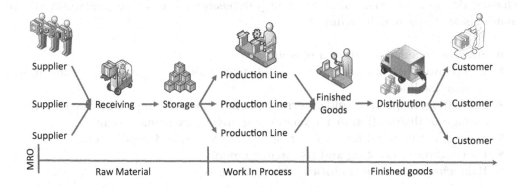

Overview

Material management is a broad topic when discussing shop floor management. It's a supply chain function that involves the sourcing, procurement, storage, and utilization of direct or indirect materials used in the manufacturing process. Direct materials are consumed during production as identified in the bill of materials (BOMs), while indirect materials are not typically used in production but are needed to support

DOI: 10.4324/b23307-18

it. Indirect materials typically include things like cleaning supplies, gloves, oil, and safety equipment. The objective of materials management is to maintain a consistent flow of material to produce quality products, on time and in the right quantity to meet customer demand.

Manufacturing material handling encompasses short-distance movement of materials inside the factory, or between the transport vehicle and building where the material is temporarily stored or shipped directly to the shop floor. This activity may require manual, semi-automated, and automated handling equipment to facility the movement of parts, boxes, crates, and pallets. Facility transport equipment typically falls into the categories of storage equipment, engineered systems, industrial trucks, and bulk material handling. Material handling can contribute to operational efficiency if material used for production is transported quickly and easily to its consumption location upon need.

Materials management prepares material replenishment plans, determines the amount of material deployed to each storage location across the facility, controls raw material inventory levels, work-in-process (WIP), and finished goods as well as communicates information regarding material needs throughout the supply chain. A fully staffed materials management team may include a Materials Manager, Inventory Control Manager, Inventory Analyst, Material Planner, Expediter, and Buyer Planner. Their focus, as a function, is to supply material, optimize inventory levels, and minimize deviations between planned and actual results.

The Materials Manager is responsible for material sourcing, purchasing, transport, receiving, inspection, storage, and inventory management. In addition, this function needs to monitor customer demand and engage in risk management. Unless a purchasing department exists, sourcing and purchasing share the responsibility of the Materials Manager which includes:

- Sourcing needed goods at a reasonable price.
- Rapid and efficient movement of goods from point of origin to an organization's storage facilities.
- Unloading, inspecting, and entering data into the company's tracking system for storage or distribution management (e.g. inventory management).
- Controlling material for turnover, spoilage, excess, and obsolescence.
- Demanding forecasting and risk management.
- Balancing material inventories with carrying costs.

Success is rooted in the five "rights" (Rs) of material management: right materials, right time, right amount, right price, and right source. In light of these, we can derive a few common material management objectives: (1) ensure continuous on-time material supply, especially with highly automated processes and a JIT (just in time) inventory model; (2) reduce material costs associated with sourcing, purchasing, transportation, and storage; (3) increase inventory turnover; (4) buy consistent quality components to ensure quality products; and (5) maintain good supplier relationships. Note that increasing turnover rates reflects how quickly your business turns over or "sells through" inventory within a given period. Higher turn rates can improve revenue and reduce the likelihood of spoilage versus a lower inventory turnover leading to increased carrying costs. In addition, everyone is in business to keep customers happy and make money. Strong relations with material suppliers build trust and cooperation toward these objectives. Make improving key supplier relationships a core focus of your manufacturing quality strategy.

Material Requirements Planning

Materials management is closely linked to shop floor activities due to its more narrow scope of continuous direct and indirect material supply relative to the logistics manager's function which includes downstream activities such as order fulfillment, product delivery, and quality assurance in transport. In addition to logistics, there is the material requirements planning (MRP) system. This is a software-based manufacturing tool designed to help the materials team understand and quantify the five "Rs" mentioned previously in terms of what materials are needed, where are they coming from, when will they arrive, and how much they will cost. These activities rely heavily on an accurate BOM in order to source the right supplies.

As with any discipline, standards should be established within the materials management function, to ensure consistent quality in sourcing, purchasing, transport, receiving, inspection, storage, and inventory management. This helps reduce process variation and maintain process stability. Receiving the right quantities of material, on time and where needed, is a hallmark of good materials management. Frequent quality checks can help to maintain system integrity along with periodic physical counts of raw material, WIP, and finished goods in inventory to detect inconsistencies or misplacements. Remember, a substantial portion of a manufacturer's expenses are for direct material costs. Therefore, inventory control is paramount to avoid tying up too much cash that can be better spent elsewhere or holding inventory too long leading to spoiled, expired, obsolete, or damaged goods.

As a side note, safety stock is a material buffer held to avoid stock-outs and continue production in case supply chain irregularities occur due to shifting supply or demand. The reorder point is the inventory set level at which an order is triggered when it's time to replenish stock. Trigger point values can be mathematically derived to ensure sufficient stock is maintained at all times.

Material Deployment and Replenishment

Materials can be transported from storage to their consumption location one or more times a shift, using people, conveyors, industrial trucks, or automated vehicles. Manual handling refers to the use of a worker's hands to move individual containers by lifting, lowering, filling, emptying, or carrying raw material and parts. People tend to be an expensive way to relocate material but provide a lot of flexibility in movement. Conveyors can be used when sequential activities occur, reducing the manual burden and cost of material transport. Multiple delivery locations may require a more expensive option for material movement such as industrial forklifts or trucks. These costs are driven by the cost of equipment and the need for human labor to operate them. Costs may be reduced through the application of automated guided vehicles (AGVs). Whenever technically and economically feasible, automated equipment can be used to reduce and sometimes replace manual material handling. Unfortunately, automated equipment is not as flexible as a human operator.

The continuous flow of raw material to a facility and WIP throughout a facility requires good planning and discipline in execution. It involves the timely packaging and transport of raw material from the supplier to the receiving location. Raw material is typically unloaded and continues with short-distance movements to its temporary storage location. When requested, material is delivered to the work area or station for consumption, transformed into finished goods, packaged and staged for transport to designated locations. Efficient flow of materials throughout a production operation is dependent on the facility layout and workstation locations. It's best to store high volume raw material closest to its consumption point to minimize movement of materials throughout the building. A material flow diagram can help visualize the proper handling and movement of in-process material between assembly areas.

Milk Run

The milk run concept in manufacturing is an in-plant material replenishment process associated with the movement of material from central storage to different locations within a facility along defined routes. It typically occurs at set frequencies, during a fixed time each day or shift, delivering small batches of materials to designated locations. The milk run driver uses replenishment orders to load bins with needed material and delivers the bins or goods, via a train-like vehicle, along a fixed route while picking up empty packing materials or bins. Material can be delivered to mini-market areas throughout the plant or directly to the production line. Modern day manufacturing facilities are using self-guided or autonomous guided vehicles as part of their material distribution systems, eliminating the delivery person from the process. These vehicles can work 24/7, often under conditions humans cannot.

Water Spiders

A water spider is the person in a warehouse or production environment who is expected to keep designated locations and workstations fully stocked with materials to maintain continuous production flow. It's a material replenishment function that allows line workers to focus on their work task without stopping to get more material. The function is intended to ensure materials are properly stocked throughout the facility while helping to isolate the waste and inefficiencies associated with material transport, allowing production line operators to concentrate on value-added work.

Water spiders are expected to follow a standardized process to support operational efficiency. Their tasks may include supplying raw materials and parts to production, removing finished goods from the work area, the movement of Kanban cards, updating status boards, and material packing, among others. Since a water spider's rounds are time-sensitive, their tasks, cycle time, and frequency of "runs" should be clearly defined and align with production needs. Operational priority is generally assigned to material replenishment, but this position may support other tasks as time permits.

The water spider will typically follow a designated route or "milk run" at a specified frequency or when triggered by material need. Properly working, the water spider can help reduce process variation and optimize material flow. A good water spider can boost productivity in support of operational excellence.

Material Handling and Storage

Material handling involves the movement, storage, protection, and control of materials and products throughout the manufacturing, warehousing, and distribution processes. In some cases, the responsibility for material handling may extend beyond material consumption and disposal. Material movement can occur manually or be aided by semi-automatic and fully-automated equipment. There are a whole host of systems available for material movement and storage ranging from conveyors to cranes, AGVs, and automated storage and retrieval systems (ASRS). There are many factors to consider when implementing or upgrading a material handling system including system performance objectives, operating conditions, ergonomic requirements, unit load (pallets, containers, or totes), space utilization, energy consumption, application of automation, existing system compatibility, and life cycle costs.

There are inherent risks in loading and unloading material from trucks, trailers, and storage shelves that need to consider employee safety and the potential for material damage. This requires that material handlers, especially forklift drivers, are trained and qualified to use designated equipment. Clear procedures and safety training should be implemented and respected by employees involved in material lifting and placement.

Tips for material handling and storage: [1,2]

- Prepare a simple and clear procedure (standard) for material handling and storage.
- Ensure area workers are properly trained and understand the material management system.
- Keep the storage area clean and organized at all times.
- Consider storing similar material together (flammable materials).
- Check incoming goods for signs of damage and mislabeling.
- Confirm the amount received is consistent with the amount ordered.
- Strive to minimize the movement of material.
- Define and adhere to storage weight limits.
- Assign addresses to storage locations for quick material retrieval.
- Ensure adequate aids are available for properly handling materials.
- Define and respect stacking limits.
- Provide sufficient space between material stacks for safe movement.
- Store heavier loads on lower shelves.
- Secure top-heavy items from tipping over or store them on their side.
- Use stair platforms to access material stored in high locations.

■ Exercise proper material handling practices when lifting and moving materials.
■ Ensure fire suppression equipment is adequate for the area based on need and risk.

Traceability

The concept of traceability can apply to raw material, parts targeted for finished goods, and production equipment such as tools and gauges. Part traceability is the ability to track every part used in production, knowing which supplier lots or serial numbers were used in a finished product, and who received the finished goods. Depending on customer and regulatory requirements, part traceability may need to include the raw materials and components used in part manufacturing and assembly, in addition to inspection results, and assembly details. To this extent, traceability can improve product quality by enhancing operational transparency for root cause analysis and corrective action when problems occur. In some industries such as medical and automotive, it's not uncommon for product traceability to include the end user, where quick recall of a product is required. Raw material and product traceability have also been extended to end-of-life disposal and recycling initiatives for certain products.

Traceability of tools and gauges can be achieved by assigning an individual identification marker such as a 2D code to each item. These markings can include plant name, shelf location number, and serial number used to manage usage. Additional information to track frequency of use and wear limits can also be collected to maintain or even extend equipment life while preserving part quality. When looking to implement a tracing system, consider a real-time system using radio-frequency identification (RFID) and bar code scanning. Ensure your traceability system has a back-up strategy and periodically confirm the continued effectiveness of the system.

Inventory Management

Inventory management is part of supply chain management. Its primary objective is to ensure the right materials are always available for production, in the right quantity and at the right time. A secondary objective is to control the level of inventory to avoid unnecessary costs due to excess stock. Other considerations include the timely sale of products to avoid spoilage, obsolescence, and the cost of materials occupying space in a warehouse or stockroom.

To manage inventories effectively, one must understand production forecasting and scheduling requirements. This demands accurate planning, robust BOM routings, effective cycle counting, timely production reporting, and good workforce and supply chain communication. A robust inventory management system strives to maintain minimum inventory levels, conducts demand forecasting, uses the First In, First Out (FIFO) process, performs regular inventory audits, prioritizes inventory management activities and maintains good supplier relationships.

Inventory can be divided into four types, raw materials/components, WIP, finished goods and maintenance, repair, and operations (MRO) supplies, all of which need to

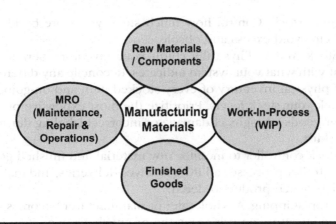

Figure 3.5.1   Categories of manufacturing materials.

be monitored and controlled. Raw material inventory consists of the materials and components needed for production. These can include food ingredients, metal, plastic, stone, and chemicals used to manufacture finished goods. WIP inventory consists of material on the manufacturing floor, past the raw material stage but at some point of maturity before becoming a finished product. Finished goods consist of inventory ready for sale and shipment. The fourth type of inventory consists of MRO inventory which includes the consumable materials, supplies, and equipment needed for manufacturing. Examples of MRO can be cleaning materials, office supplies, tooling, lab equipment, personal protective equipment (PPE), and much more. It's important that MRO materials are not only available in the facility, but they are also quick and easy to acquire, when needed. Figure 3.5.1 displays the four categories of manufacturing materials.

Real-time tracking of materials plays a key role in inventory management. Concepts like plan for every part (PFEP) can be employed for end-to-end inventory optimization to reduce shortages, improve working capital, and maximize production throughput. The focus of PFEP is to right-size inventory stock to avoid the threat of a part shortage impacting production.

The implementation of standard processes and data analysis tools can help establish enterprise resource planning (ERP) system stability and longevity that can be used to proactively optimize the supply chain. Despite regular maintenance needed to maintain ERP system stability, critical issues can disrupt daily operations such as expedited purchase orders (POs), part shortages, shifts in customer demand, and changes to the sales forecast. These need to be managed on a case-by-case basis. [3] The following are several tips for inventory management.

Tips for Inventory Management [4,5]

- Ensure accurate forecasting. Consider historical sales, market trends, predicted growth, the economy, promotions, and marketing activities.
- Practice FIFO. This requires selling goods in the order they are made to avoid spoilage, obsolescence, or damage.

- Manage aging stock. Control how much stock you store by identifying limits. This can help avoid excess and obsolescence.
- Perform Stock Audits. Physically count your inventory periodically, matching your count with what your system indicates. Reconcile any differences. Consider an annual physical inventory of every stocked item and on-going spot-checks.
- Always track your stock levels. Prioritize the most expensive products.
- Check your goods for signs of damage or incorrect labeling during stock audits. Use a checklist.
- Assign a stock controller to monitor raw material and finished goods inventory. A stock controller processes all POs, receives deliveries, and makes sure incoming materials match product ordered.
- Consider drop shipping. A wholesaler or manufacturer becomes responsible for carrying the inventory and shipping the products when an item is ordered. This avoids the worry of inventory holding, storage, or fulfillment.
- Engineer material flow into the process, from raw to finished goods, and from one process to the next.
- There are several different technologies to help with material tracking, control and maintaining operational efficiency including Barcoding, RFID and the internet of things (IoT). Consider these to enhance material control.

Health and Safety

Health and safety are important in the movement of material. Employees can suffer musculoskeletal disorders due to sprains and strains to the lower back, shoulders, and upper limbs. Exercising good practices in lifting and limiting the weight of materials movement can help alleviate these afflictions. Make ergonomic and proper lifting practices an integral part of your processes to avoid chronic pain, disability, medical treatment, and financial stress that can impact workers.

Ergonomic interventions can play a key role in lowering the incidence and severity of worker injury while improving a company's productivity, quality, and overall business image. Good practices can include minimizing manual lifting tasks and using positioning equipment like lift/tilt/turn tables, hoists, balancers, and manipulators to minimize human reaching and bending. Take a good look at how best to use individual energy, available equipment, and personal exertion to get the job done in the most efficient and effective way possible.

SIDEBAR: MATERIAL KITTING

When managing a multitude of customized products, product variants with small components, or a lack of adequate shop floor space, kitting of components may be an option. Kitting involves the bundling of BOM components in a packet which is available to operators for product assembly. This approach can potentially simplify line-side inventory, increase assembly speed and improve the quality of assembly.

Key Points

- Prepare guidelines and standards for proper material handling and storage. Follow them!
- Ensure people are properly trained in material handling to minimize personal injury and material damage.
- Periodically confirm your traceability system is working as designed.
- Conduct periodic cycle counts to confirm the physical inventory counted matches inventory records. If not, investigate and reconcile!
- Consider material "kitting" when a lot of small components or lack of adequate working space is available.
- Remember the five "rights" of material management, the right materials, right time, right amount, right price, and right source.

Sources

[1] https://www.mhi.org/fundamentals/material-handling
[2] https://www.flexqube.com/news/guide-basics-successful-material-handling/
[3] https://www.leandna.com/blog/tuneup-erp-plan-every-part-pfep/.
[4] https://www.hashmicro.com/blog/inventory-management-hacks-for-manufacturers/
[5] https://www.bluecart.com/blog/manufacturing-inventory.

3.6

Manufacturing Equipment, Fixtures, and Tooling

It matters little how much equipment we use; it matters much that we be masters of all we do use.

~ Sam Abell

Objective: Maintain equipment, fixtures, and tooling in good working condition and continuously strive to improve their reliability and robustness over time.

Overview

Manufacturing equipment, fixtures, and tooling aid in the conversion of raw materials to finished goods through mechanical and chemical processes. Their availability depends on needed utilities such as electricity, gas, air, and a skilled workforce to effectively use them. The use of machines and tools also play a key role in manufacturing profitability since organization capital is often frozen in these company assets along with the expense of maintenance. An efficient and effective manufacturing operation requires that equipment, fixtures, and tooling be continuously functional, clean, easily identifiable, and readily available. This comes down to having a good maintenance program and equipment traceability system for regularly stored items. Some equipment may require periodic adjustments or maintenance due to their frequent use, the working environment, or overall complexity. Properly maintained equipment can often last well beyond their design life.

Preventive maintenance, according to the manufacturer's recommendations, and incremental improvements in equipment robustness, based on actual use conditions, can help realize or extend life expectancy. However, maintaining equipment availability or extending a machine's lifespan comes with a price and requires a disciplined mindset to schedule and execute maintenance on time while justifying the expense for equipment upgrades. Let's explore the world of equipment, fixture, and tooling management in more detail.

DOI: 10.4324/b23307-19

Equipment Optimization (Effectiveness, Reliability, and Durability)

Wear & tear will likely impact equipment efficiency over time. It takes thought, intention, and action to maintain and keep equipment running at peak performance. Equipment effectiveness, reliability, and durability are aspects of equipment management that will influence manufacturing performance. Done right, good equipment management will lead to optimized equipment availability and performance. Optimizing manufacturing equipment requires a clear plan and the discipline to follow that plan, regardless of manufacturing hiccups or constraints. That's not to say that manageable deviations from the plan can't occur, it simply means that any deviation from plan must be justified and efforts taken to get the plan back on track as soon as possible.

Effective shop floor management requires that critical operating machines, tools, and fixtures be properly maintained, monitored, and controlled to avoid unplanned downtime or degradation in performance. The objective is to maintain equipment at peak performance to produce quality parts when needed. Any deviation from planned equipment availability or performance will impact manufacturing productivity. Overall equipment effectiveness is an industry standard for measuring manufacturing productivity. This equipment effectiveness metric is often complemented by a robust total productive maintenance (TPM) program designed to minimize, if not prevent, unexpected equipment failures. Preventing equipment failures through good TPM practices helps to maintain equipment effectiveness and manufacturing productivity. In the following pages, we will focus on these two key contributors to equipment effectiveness, reliability, and durability along with a few other factors influencing equipment, fixture, and tooling performance. Let's start with equipment effectiveness and the topic of OEE.

Overall Equipment Effectiveness

OEE is a standard and best practice for measuring manufacturing productivity. It's the percentage of manufacturing time that is truly productive. An OEE score of 100% means you are manufacturing all good parts, as fast as possible, without interruption. There are three main factors that contribute to an OEE score of 100%, they are 100% quality in which only good parts are produced, 100% performance when parts are produced as fast as possible, and 100% availability when no downtime occurs during scheduled production.

Availability considers the duration of *planned* and *unplanned* production-line stops. It accounts for losses caused by events that stop planned production for a significant length of time. An availability value of 100% means the process is always running during planned production time. Performance considers speed losses (relative to maximum speed) due to *Slow Cycles* and *Minor Stops*. The performance value is determined by comparing Actual Cycle Time (or Actual Run Rate) to ideal cycle time (or ideal run rate). A performance value of 100% means when running, the process is operating at maximum capacity. Quality considers parts which do not meet quality requirements (defects) and parts that require rework. A quality value of 100% means there are no defects, only good parts are being produced. OEE is a good metric for identifying efficiency losses, benchmarking progress, and improving the productivity of manufacturing equipment through waste avoidance and elimination. Let's consider the topic of efficiency losses in more detail.

Efficiency Losses

Efficiency losses can be looked at from the perspective of availability, performance, and quality losses. These losses are often discussed in terms of the six common causes of equipment-based productivity losses (six big losses) in manufacturing. They include:

Availability Losses

- **Unplanned Stops** – any line stop in which equipment scheduled for production is not running due to issues such as equipment or tooling failure, no operators or lack of material.
- **Planned Stops** – any significant time periods in which equipment is scheduled for production but is not running due to a changeover or other equipment adjustments. Examples include setup, changeovers, major adjustments, and tooling adjustments. This category could also include cleaning, warmup time, planned maintenance, and quality inspections.

Performance Losses

- **Minor Stops/Idling** – any time where the equipment stops for a short period of time (two minutes or less). These stops are typically chronic problems that can be corrected by the operator and don't require maintenance personnel. Examples include misfeeds, material jams, obstructed product flow, incorrect settings, misaligned or blocked sensors, equipment design-related issues, and periodic quick cleaning.
- **Slow Cycles (Reduced Speed)** – any time equipment runs slower than its *ideal cycle time* which is the fastest possible time to manufacture one part. Contributors to reduced speed include dirty or worn equipment, poor lubrication, substandard materials, poor environmental conditions, operator inexperience, startup, and shutdown. This category includes anything that prevents the process from running at its maximum possible speed.

Quality Losses

- **Production Rejects** – any defective parts produced during stable (steady-state) production. This includes scrapped and reworked parts, since OEE measures First Pass Yield quality. Causes of process defects can include incorrect equipment settings and operator or equipment-handling errors.
- **Start-up Rejects (Reduced Yield)** – any defective parts produced from startup until stable (steady-state) production is reached. This includes scrapped and reworked parts which can occur after equipment startup or changeovers. Contributors include suboptimal changeovers, incorrect settings when a new part is run, equipment that requires warmup cycles, or equipment that inherently creates waste after startup.

OEE Factors Losses	Six Big Losses	Description	Next Level	Examples
Availability Losses	Unplanned Stops	Equipment downtime (Breakdowns)	Maintenance	Unplanned or longer than planned
			Machine / Equipment	Equipment failure
			Material	Wrong material, Material shortage
			Non categorized downtime	Default value set by system
			Personnel	Labor shortage
			Quality	Jidoka stop
			Set-up / Changeover Delay	Changeover longer than planned
			Technical	Downtime due to SMT Equipment (e.g. printer, pick & place)
	Planned Stops	Set-up & Adjustments	Material	Replenish
			Personnel	Meeting, Training, Shift team exchange
			Equipment	Planned maintenance
			Quality	Changeover quality check
			Set-up / Changeover	Start-up, warm-up, cleaning
			Technical	Machine upgrade
Performance Losses	Minor (small) Stops	Idling	Misfeeds, material jams, obstructed product flow, dropped product, incorrect settings, misaligned or blocked sensors, cleaning	
	Slow Cycles	Reduced speed	Dirty or worn equipment, poor lubrication, sub-standard materials, poor environmental conditions, operator inexperience or shortage but line producing, rough running, startup, & shutdown	
Quality Losses	Production Rejects	Scrap & rework	Incorrect equipment settings, operator or equipment handling errors, product out-of-spec, damaged product	
	Start-up Rejects	Reduced yield	Suboptimal changeovers, incorrect settings (when a new part is run), incorrect assembly, equipment requiring warmup cycles, or equipment that inherently creates waste after startup	

Figure 3.6.1   OEE losses and examples.

The difference between equipment failure (availability loss) and a minor stop (performance loss) is typically defined as two minutes. Any stop 2 minutes or less is a minor stop. Any stop greater than 2 minutes must have a reason and is considered equipment failure. An overview of the OEE losses can be found in Figure 3.6.1.

A key objective of TPM is to help reduce, if not prevent, the six big losses from significantly impacting equipment-based productivity performance. Let's turn our attention to calculating OEE.

OEE Calculation

The formula for OEE calculation is as follows:

$$\text{OEE} = \text{Availability} \times \text{Performance} \times \text{Quality}$$

$$\text{OEE} = \frac{\text{Run Time}}{\text{Planned Production Time}} \times \frac{(\text{Ideal Cycle Time} \times \text{Total Parts})}{\text{Run Time}} \times \frac{\text{Good Parts}}{\text{Total Parts}}$$

Descriptions

- Availability = run time / planned production time
- Performance = (ideal cycle time × total parts) / run time
- Quality = good parts / total parts
- Run Time = Planned Production Time (PPT) − Stop Time = (PPT − Stop Time)

Availability considers all events that stop planned production long enough (>2 minutes) to assign a known cause. Availability is calculated as the ratio of *Run Time* to *Planned Production Time* as follows:

Availability = Run Time / Planned Production Time = (Planned Production Time − Stop Time) / (All Time − Schedule Losses)

where:

- Run time = (Planned production time − stop time).
- Planned production time − total time that equipment is expected to produce. Calculated by subtracting *Schedule Losses* from *"All Time"*.
- *All Time* − every minute of every day − 24 hours per day / 7 days a week.
- *Schedule Losses* − losses to consider are *mandatory breaks, defined lunch time, exercise time and scheduled meetings.*
- *Stop Time* − all the time where the manufacturing process was intended to be running but was not, due to Unplanned Stops (e.g., Breakdowns) or Planned Stops (e.g., Changeovers) which are considered manufacturing process downtime!

Performance considers anything that causes the manufacturing process to run at less than maximum possible speed. Performance is the ratio of *Net Run Time* to *Run Time* and is calculated as:

$$Performance = \left(Ideal\ Cycle\ Time \times Total\ Parts \right) / Run\ Time$$

where:

- *Net Run Time = Ideal Cycle Time × Total Parts* or the fastest possible time to manufacture parts.
- *Ideal Cycle Time* − the fastest cycle time that your process can achieve under optimal conditions. Best possible cycle time: no other factors are considered.
- *Total Parts* − Total shippable parts (including reworked and inspected. Scrapped parts are <u>not</u> included).
- *Run Time* − is calculated by subtracting stop time (e.g., manufacturing process downtime) from planned production time. Run time includes time when the process could be experiencing small stops, reduced speed, and making reject parts.

Performance should never be greater than 100%. If it is, that usually indicates that *Ideal Cycle Time* is not correct (too high).

Quality considers manufactured parts that do not meet quality standards, including parts that require rework. Quality is calculated as:

$$Quality = Good\ Parts\ /\ Total\ Parts.$$

where:

■ *Total Parts* – Total shippable parts including reworked and inspected (scrapped parts are <u>not</u> included).
■ *Good Parts* – parts that meet quality standards (e.g., units that pass through the manufacturing process the first time without needing rework).

In the calculation, a single number captures the production line OEE, while the three components of *availability, performance, and quality* highlight the fundamental nature of operational losses. The OEE value provides an accurate picture of how effectively your manufacturing process is running, while making it easy to track process improvements over time. Unfortunately, it does not provide the underlying causes of lost productivity.

OEE is commonly applied to discrete manufacturing processes; processes that make individual parts. However, it can also be applied to continuous processes (refineries). This is because OEE identifies the ratio of Fully Productive Time (actual output) to Planned Production Time (theoretically possible output). The difference between the two is lost time (waste) that could otherwise be used for manufacturing.

OEE Improvement Cycle (with Plan-Do-Check-Act [PDCA])

OEE provides a standard method for benchmarking and driving progress improvement by highlighting problems for root cause analysis and elimination. By measuring OEE and corresponding losses, you gain insights on how to improve the manufacturing process. One of the fastest ways to improve is to focus on the top losses. Consider applying the PDCA cycle for improving OEE performance.

Plan – Collect loss data. A top losses report or Pareto chart can identify the largest sources of lost production time. Determine the biggest loss factor (e.g., downtime, setup, adjustments, scrap, rework, etc.)

Do – Analyze the loss factor chosen for root cause analysis. Identify a countermeasure. Focus on the inputs that drive output results. Implement the countermeasure. Make sure every countermeasure has an "owner" and due date.

Check – Verify effectiveness. Capture lessons learned.

Act – Determine next steps. Implement actions to affect change. Update standard work instruction to reflect best practices. Display your results. Show the impact of your actions to drive improvement.

Equipment Improvement Metrics (MTTF and MTBF)

Mean time to repair (MTTR) is a basic measure of maintainability. It represents the *average time* required to repair a failed machine or piece of equipment; the actual time working on a machine repair. The average time to repair (MTTR) is calculated as follows:

$$MTTR = \frac{\sum \left(\text{Total maintenance time} \right)}{\sum \left(\text{Number of repair events} \right)}$$

Total maintenance time (reaction time + repair time) is the time when a machine or piece of equipment is not available for scheduled production due to an unexpected breakdown. Reaction time is the time from when the equipment was identified as down (failed) to the start of equipment or machine repair. Repair time is the time taken to work on the machine repair. The repair time starts from the time the line stopped due to equipment breakdown (failure) until the time the production line was approved or available for making good parts. An example for calculating MTTR is as follows: if you spent 15 hours of unplanned maintenance on an asset that failed three times during a month, the MTTR is 5 hours.

Mean time between failures (MTBF) is the predicted elapsed time between inherent failures or downtime of a machine or equipment during normal system operation. It's the *average time* to run a line or machine without failure. MTBF can be calculated as follows:

$$MTBF = \frac{\sum \left(\textit{Number of machine operating hours} \right)}{\sum \left(\textit{Number of machine downtime events} \right)}$$

The higher the MTBF, the longer a system is likely to function before failing. For example, if a machine was operating for 120 hours in a month and during that time the machine broke down two times, the MTBF for the asset is 60 hours. A visual of the key parameters discussed can be found in Figure 3.6.2.

MTTR and MTBF are key metrics for evaluating equipment availability. They describe machine reliability, effectiveness of maintenance and responsiveness to machine breakdowns, and repair. MTTR and MTBF are machine indicators. Although reported by line, they are not line indicators of efficiency, they are tools used to compare machines within a facility and identify where Engineering/Maintenance efforts should focus. Improvement in the MTTR and MTBF metrics will increase equipment availability and OEE performance.

Spare Parts Management

Sometimes it's difficult to get timely replacement parts for equipment repair. This can have a significant impact on manufacturing productivity when equipment failure occurs unexpectedly, especially during peak periods of equipment demand.

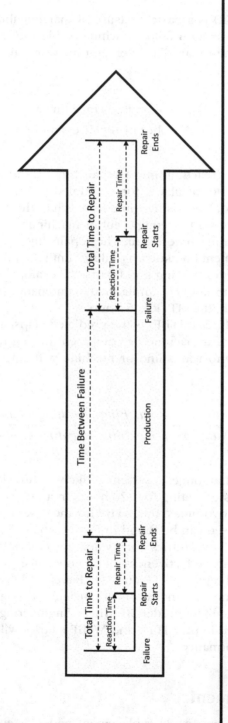

Figure 3.6.2 Visual of Time to Repair and Time between Failure.

In response to the unexpected downtime events, plants can evaluate their key operating equipment and highlight which spare parts are difficult to acquire in a timely manner. By identifying their *critical spare parts* and working to maintain 100% of critical spare parts in inventory, a facility can mitigate their risk of extensive equipment downtime caused by waiting for replacement parts to arrive.

Unfortunately, it may be costly to maintain an inventory of spare parts, but this cost should be compared to the potential impact of unplanned downtime. In some situations, parts can be machined, or 3D printed in house, helping to reduce inventory costs and equipment downtime. Spare parts management becomes more efficient when multiple machines of the same type are used at a location.

It's recommended that a critical spare parts list be prepared and managed. It can be updated whenever lack of a part in inventory causes unexpected and significant equipment downtime. Plants should consider preparing a process for managing their spare parts inventory and audit it periodically for effectiveness in avoiding excessive downtime. Operationally excellent plants prepare clear procedures for managing their spare parts inventory and audit the inventory periodically to confirm critical part availability.

SIDEBAR: TIPS FOR REDUCING EQUIPMENT DOWNTIME – GOOD OPERATING DISCIPLINES

- **Triggers** – implement process triggers (visual controls) to highlight failures quickly!
- **Escalation** – define a clear escalation procedure that explains what to do if something breaks; whom to call, how to document what happened, and how to start the repair process.
- **Problem-Solving** – structure your approach for trouble shooting and equipment repair.
- **Accountability** – establish clear repair ownership and set expectations for the repair.
- **Follow-up** – monitor for timely execution of the repair.
- **Training** – teach problem-solving and cross-train employees to create multiple-response resources.
- **Failure Mode and Effects Analysis (FMEA)** – use an FMEA or *Fault Tree Analysis* to capture lessons learned and continuously update existing trouble shooting guidelines for known and commonly occurring issues.
- **Postmortem / Incident Retrospective** – use your understanding of root cause to implement actions that *prevent* reoccurrence. Update TPM practices with lessons learned. Deploy improvements across the shop floor where applicable!
- **Spare Parts Inventory** – maintain an inventory of long-lead time, common and critical parts needed for equipment repairs.
- **Equipment supplier support** – expect and negotiate 24/7 support from key suppliers in response to specialized equipment that may need immediate maintenance.

Equipment Reliability

Reliability is evident when equipment performs its intended function consistently well and without failure. Reliability can be designed into a product but is more often realized through the application of lessons learned from equipment maintenance and repair. This is typical of today's automobiles. Cars are much more reliable now than they were 10 or 20 years ago, as evident by the decrease in unexpected breakdowns. However, they will not perform without fail if activities like regular oil changes and scheduled maintenance routines are not completed at recommended intervals. Certain equipment, such as measurement devices, may require periodic calibration and analysis for their continued ability to produce repeatable and reproducible measurements at acceptable levels of accuracy and precision.

The shop floor needs reliable equipment. This reliability will result from the type of equipment chosen, its application environment, equipment-handling practices, calibration frequency, and performance of regular maintenance. Deployment of a robust TPM program will help to maintain equipment integrity and availability over its expected lifespan which may improve with the application of lessons learned. The following topics will touch on some of the key activities that influence equipment reliability.

Equipment Calibration

Equipment accuracy and precision may tend to degrade over time with repeated use. As needed, a periodic check of an instrument's continued accuracy and precision is recommended and often specified by the original equipment manufacturer as part of an on-going equipment maintenance program. This is common practice with measurement equipment.

If a significant deviation from the original equipment performance standard is detected, calibration of the equipment will likely be required. Equipment calibration is done to a set of known reference standards used to bring an instrument back into alignment with a standard or in compliance with its original design parameters. The objective of calibration is to minimize measurement or output uncertainty to an acceptable level. In the specific case of measurement equipment, calibration activities should ensure traceability to a reference standard. The following are some tips for managing equipment calibration.

Tips for Equipment Calibration

- Maintain a list of equipment requiring calibration. Record the frequency of required calibration along with other identifying information.
- Designate someone to take responsibility for equipment calibration.
- Frequently audit equipment for calibration compliance to verify system integrity.
- Review the status of equipment calibration as a periodic agenda item in a management review.
- Define a reaction plan for equipment found out of calibration.
- Keep a historical log of calibration activities for reference.

Measurement Systems Analysis

Measurement systems analysis (MSA) seeks to identify the components of variation in a measurement system. Measurement system integrity is essential for ensuring consistent and conforming product is being delivered to customers. Analysis of a measurement system starts with equipment calibration and is generally followed by assessing measurement gauge repeatability and reproducibility, commonly known as GR&R. GR&R is the process of evaluating an instrument's precision to ensure its measurements are repeatable and reproducible within specific tolerances. For example, in the automotive industry, equipment GR&R is expected to contribute no more than 15% of overall process variation. Anything higher needs to be addressed with improvement actions.

Measurement system variation can be characterized using various techniques to understand measurement accuracy (e.g. bias, stability, and linearity) and precision (repeatability and reproducibility) relative to a reference or standard. MSA is used in manufacturing for multiple purposes such as evaluating new measurement equipment capability, comparing measurement devices, as a baseline for improving measurement instruments and as a process variation acceptance criterion. It's often used in the transfer of equipment from one location to the next, ensuring equipment integrity is maintained during the move.

The "repeatability" component of GR&R looks at gauge-induced variation in measurements by having one operator measure the same part characteristic several times using the same measurement instrument. Reproducibility considers operator-induced variation and other sources. It's performed by having different operators, using the same measurement equipment, measure the same part characteristic multiple times. These results are then evaluated relative to overall process variation.

There are three types of GR&R studies: crossed, nested, and expanded. The type of technique used will depend on data availability and the type of measurement testing performed (destructive or non-destructive). An overview of MSA can be found in Figure 3.6.3. More can be found on this topic from the source reference [1].

Total Productive Maintenance

TPM is key to maintaining equipment reliability. It's intended to sustain OEE by preventing equipment deterioration, while facilitating an environment of ownership between operators and their machines. A good TPM program leverages lessons learned, best practices, and standard work procedures to improve maintenance efficiency and effectiveness. Lessons learned should be considered in future equipment design to create more robust and reliable equipment.

TPM can provide a comprehensive approach to equipment management, increasing equipment reliability and availability for production. A well-executed program can significantly reduce equipment failures, avoid production defects, and prevent accidents. TPM has five major elements including *equipment cleanliness*, typically pursued as part of a 5S initiative, *routine maintenance* intended to maintain design functionality of equipment, *improvement maintenance* focused on reducing the occurrence of breakdowns, *predictive maintenance* used to anticipate breakdowns,

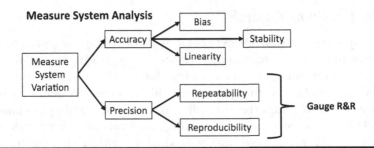

Figure 3.6.3 Measurement system analysis overview.

and *preventive maintenance* to avoid equipment breakdown. This equipment maintenance strategy is needed to support a just-in-time production system for achieving continuous flow and process optimization.

Periodic TPM audits should be performed to ensure the system for equipment maintenance is on schedule and effective. It's recommended that a multi-skilled and cross-functional group of experienced employees performs these audits, looking for system issues with training, schedule execution, maintenance practices, and improvement opportunities.

We discussed TPM from an operator responsibility perspective in the last chapter. In this chapter, we will touch on TPM from an equipment maintenance perspective. A good TPM program has several key elements including reducing the impact of the six major OEE losses (discussed previously), autonomous maintenance, systemic maintenance, training and qualification, and asset management. Let's review these items in more detail.

Autonomous Maintenance

The concept behind autonomous maintenance is to develop operators to independently execute low- to mid-level maintenance tasks and basic repairs for optimal equipment performance. This frees up the experienced maintenance team to focus on higher value-added activities and complex repairs. This approach requires several steps including operator training to maintain equipment within their scope of work. This should include knowing the requirements for scheduled equipment maintenance consisting of cleaning, lubrication, and worn part replacement. Activities such as dust and dirt removal, tightening of nuts and bolts, expectations for oiling and lubrication, and recognizing the signs of wear and tear, may also be part of the training. One of the autonomous maintenance goals is to maintain equipment in optimal working condition through good housekeeping practices. To do this safely, lockout/tagout (LOTO) procedures should be a required part of the training.

Implementing standard practices in equipment maintenance, rooted in the equipment manufacturer's specifications and recommendations are encouraged. Clear communication on what tasks to do, how to do them, and how frequently they should be done is important. Checklists can be an effective tool in this regard. Inspections and monitoring are also good practices for autonomous maintenance. Operators should periodically inspect their equipment and provide feedback whenever their equipment's condition appears to be substandard. Visual aids can be used to reinforce expectations and help operators follow standards for equipment and machine maintenance. Finally, encourage operators to suggest improvements in how autonomous maintenance activities are performed. Continuous improvement is a hallmark of operational excellence. The ultimate objective is to prevent equipment deterioration and keep operations running at a stable and predictable rate of performance. The following are seven steps for autonomous maintenance:

1. Enhance operator knowledge based on expectations.
2. Clean and inspect equipment. Restore equipment to operating standards.
3. Reduce causes and eliminate sources of equipment contaminants.
4. Define standards for cleaning, maintenance and inspection.
5. Monitor performance and conduct periodic inspections.
6. Visualize maintenance management KPIs.
7. Drive continuous improvements.

Systemic Maintenance

Systemic maintenance is exactly as it sounds: create a logical equipment maintenance plan that takes a disciplined approach to all maintenance activities. This requires a thorough understanding of the maintenance requirements for each piece of equipment followed by preparation of standards for condition-based maintenance tasks with assigned ownership for equipment operators and the maintenance staff. Operators focus on equipment care while the maintenance staff develops practices and procedures targeted at preventive maintenance activities. Maintenance staff may also be expected to train, coach, and support new operators in completing their tasks while tackling the more complex problems encountered with equipment breakdowns. The ultimate objective is zero equipment breakdowns and zero defects due to equipment.

Training and Qualification

The development of people skills and expertise is needed to preserve and improve machine functionality. This includes an understanding of the "know-how" and "know-why" of equipment maintenance. The goal is to elevate the skills and abilities of all people working on the shop floor in lean manufacturing and preventive maintenance. In larger facilities, employee development can lead to a certification program to become a manufacturing operations technician.

Asset Management

Asset Management is about maintaining and improving manufacturing equipment to minimize their operating cost. It involves taking a systematic and cost-effective approach to developing, operating, maintaining, upgrading, and disposing assets. This is best accomplished with a comprehensive plan that is viewed periodically for effective deployment, proper controls, and verification of effectiveness. If you recognized the value of properly maintaining your key capital assets but have not initiated a TPM program yet, the following steps can help you get started.

Implementing TPM

1. Select a pilot area to start.
2. Return equipment to its original operating condition (Reference the equipment manual).
3. Start measuring OEE (for baseline performance).
4. Identify and work to reduce the six major losses.
5. Prepare and implement planned maintenance.
6. Schedule audits to verify compliance and effectiveness.

Equipment Durability

Equipment durability is the ability to withstand wear, tear, deterioration, and break-age under normal operating conditions and throughout the equipment's design life-time, without excessive maintenance or repair. Durability can be assessed relative to several measures including usable hours, lifespan, and the number of operating cycles. It can be predicted based on equipment maintenance and repairability. Equipment's physical design properties should also be considered when assessing durability including resistance to dust, temperature, rusting, rotting, and fatigue, in addition to its constructed material toughness and aging properties. When evaluating new equipment for purchase and installation, consider these factors in your selection criteria and decision-making process.

Improving the durability of existing and new equipment can be achieved through the application of lessons learned. Knowledge acquired about equipment design, wear and tear can become valuable information used to make minor, and some-time major, improvements in components that can reduce breakdowns and extend equipment life. One example is a material dispense mixing spindle that needed to be replaced often, due to excessive wear, resulting in frequent equipment downtime and replacement cost. When a technician realized that he could replace the plastic material by machining a more durable stainless-steel version of the component, the part cost increased slightly but lasted considerably longer, significantly reducing replacement cost, unnecessary downtime for maintenance, and improved dispense machine robustness. In short, valuable information can be gained by using observation, experience, and a little foresight, to make incremental improvements in equipment components and design, leading to enhanced durability.

Equipment maintenance reveals a lot about how equipment works and performs under certain application conditions. Environmental cleanliness, equipment design, operating time, maintenance practices, application methods, and the like, all play a role in the availability, performance, and lifecycle of a machine. While engaging in equipment maintenance and repair, it's important to note design weaknesses and improvement opportunities as a form of lessons learned for continuous improvement. Fastening a sensor in place to avoid its movement or changing a machine component's base material from plastic to steel all help to incrementally improve certain aspects of equipment performance, robustness, and durability. Consider enhancing your preventive maintenance routines through lessons learned and continuously updating maintenance work instructors to capture and share improvements with others.

Key Points

- Preventive maintenance practices reduce unplanned equipment and production line downtime.
- Use OEE, MTTR, and MTBF to monitor and improve manufacturing productivity.
- Skipping preventive maintenance routines can lead to premature and unplanned equipment failures as can be observed by a lack of car maintenance.
- Focus on developing a "cycle" versus "time-based" preventive maintenance schedule since this will be more reflective of actual machine usage.
- Improving OEE performance requires moving from an unplanned to a planned maintenance schedule and working to reduce the major equipment availability losses.
- Critical spare parts should be held in inventory to reduce repair time.
- A critical spare parts list should be established and updated whenever a part, not in inventory, causes unexpected and significant equipment downtime.
- MTTR indicates the effort to bring a machine back into production, whereas MTBF is an indication of how often machine failure occurs.

Source

[1] https://asq.org

3.7

Methods

Let your intentions create your methods and not the other way around.
~ Peter McWilliams

Objective: Define shop floor practices and procedures that minimize process variation to obtain consistent and predictable output results while continuing to make productivity improvements.

Overview

A method is a structured approach for accomplishing something. It often reflects a logical and systematic way of executing a series of activities or a process, in a planned and orderly way, to achieve a specific result. Manufacturing methods put into practice what an organization wants to do, how it wants to do it, and, to a certain extent, the behaviors it expects of the workforce in doing it. Typical examples of shop floor methods include:

- Purchasing materials.
- Product assembly.
- Material handling.
- Data gathering and display.
- Conducting Gemba walks.
- Capturing and closing action items.
- Production scheduling.
- Change management.
- Problem-solving.

There are an infinite number of methods, some well-defined and accepted as industry standards, while others reflect the unique needs and expectations of an organization

DOI: 10.4324/b23307-20

in satisfying its goals and objectives. Let's take time to review some of the more common methods used in daily shop floor management, starting with the plan-do-check-act (PDCA) cycle, which is familiar to many people.

Plan-Do-Check-Act

PDCA is an iterative method used in manufacturing for process and product control as well as continuous operational improvement. It's sometimes called the Deming cycle and has often been described as the PDCA or plan-do-study-act (PDSA) cycle. The primary objective of this method is to acknowledge the current state, prepare an improvement plan, execute the plan, gather data and information, and check the results against expectations. If needed, additional actions may be taken to achieve the desired state or drive continuous improvements.

This simple method can be very effective when applied with clear intent and discipline. The following outlines each step in brief.

- **Plan** – assess the situation that needs to be addressed. Prepare an action plan.
- **Do** – execute the plan, gather data and information for evaluation.
- **Check/Study** – compare the results to expectations. Analyze the data and identify gaps.
- **Act** – make adjustments or take corrective action. Start working to the new/updated standards.

5S Method

As in the case of PDCA, 5S is a fundamental "housekeeping" method for all employees to embrace and exercise. It's divided into five phases: Sort, Set in Order, Shine, Standardize, and Sustain. These phases are intended to create a clean, organized, efficient, and effective workplace. The objective is to eliminate what's not needed from the workplace, ensure what's needed is clean and working properly. The next step is to standardize these practices, so they are sustainable over time. There are many different sources available where one can learn more about how to implement and maintain the disciplines of 5S. However, what's important is that a shop floor operation has a clear and structured approach to workplace cleanliness and orderliness that creates a respectable, efficient and comfortable working environment. The following is a brief overview of each phase of the 5S methodology.

- **Sort** – this activity focuses on sorting through workplace items, keeping what's needed and discarding what's not. This helps increase workspace, decrease work distractions, and eliminate the time wasted looking for needed items.
- **Set in order** – this activity is concerned with arranging and labeling necessary items for visibility, easy access, and use.
- **Shine** – this activity requires keeping a clean and well-maintained workplace, ensuring all equipment is working properly.

- **Standardize** – this requires putting a process in place to ensure the practices of sort, set in order and shine are continuously exercised and respected.
- **Sustain** – this requires periodic verification that the 5S methodology continues to work, as designed, with corrections to deviations made and training performed as needed.

5S Tips

- Make 5S an inherent part of the company culture.
- Introduce new employees to 5S during the on-boarding process.
- Assess 5S compliance during Gemba Walks.
- Include 5S as part of layered process audits.
- Recognize and reward people for their 5S discipline, behavior, and compliance.

Deviation Management

One of the most important shop floor activities needed to establish and maintain process stability and control is deviation management. A deviation is an ***unexpected*** or ***undesirable*** change of a process ***state or characteristic*** away from its specified standard, norm, or desired target. Process and system deviations can cause disruptions to workflow, leading to process instability and result in product defects, if not identified and addressed quickly. Deviation management is a relatively simple concept that leverages visual clues and operating parameters to identify deviations from process standards and targets. The cause of deviations must be understood and action taken, if needed, to return the process back to its expected level of performance.

Deviation management is as integral to daily shop floor management as monitoring the gas gauge while driving your vehicle. If you ignore the critical parameters required to maintain operational stability, your process will start to deteriorate, generate defects, and eventually lead to customer dissatisfaction when resulting products no long meet expectations. To prevent this from occurring, everyone in the organization must practice deviation management by highlighting and addressing abnormalities discovered during their daily work routines. It must become an inherent part of the company's operating practices, from the line operator to the company CEO since deviations can occur at any level and within any process of the organizational hierarchy. Executing the disciplines of deviation management and holding employees accountable to practicing it continuously help reinforce operational stability and the behaviors required for manufacturing excellence.

Deviations can be classified into three categories: deviations in performance, process, and system execution. Performance issues are people centric and can typically be addressed immediately, reinforced through coaching and on-the-spot procedural training. Deviations associated with process issues are generally unique to a manufacturing area or operation and should be managed by the area team leader. Systemic issues tend to impact a broader scope of operations, affecting more than one process, area, or function in the plant. Core process and systemic deviations may take more time to understand the root cause, along with implementing an effective corrective action to eliminate their reoccurrence. To effectively deploy deviation management, employees should consider the following activities:

- **Identify** – monitor the process. When executing daily work routines, employees should continuously monitor the process for abnormalities and deviations from process work instructions, operating standards, and manufacturing targets.
- **Investigate** – investigate abnormalities and deviations. The issue owner must take time to understand the causes of process abnormalities and deviations. Understanding a cause will help team members identify effective solutions to eliminate process disruptions.
- **Correct** – implement actions to eliminate a deviation's root cause. If more time is needed to understand a problem, put a containment in place which allows production to continue, until root cause can be properly identified and eliminated. Corrective actions deployed should prevent reoccurrence since eliminating the same deviation more than once is waste.
- **Follow up** – confirm effectiveness of actions. Upon implementing a process change or correction, best practice is to follow up with the change to confirm continued compliance and effectiveness of actions taken. If effectiveness is questioned, consider if the true root cause was identified prior to solution selection and implementation.

Deviation Management Tips

- Look for significant process deviations every day, especially during Gemba walks.
- Correct performance issues immediately, when possible.
- Use visual controls to highlight deviations and trigger actions.
- Capture deviations as actions to be prioritized and addressed.
- Consider following up with the status of open actions (unresolved deviations) during performance reviews.
- Use layered process audits to verify conformance to standards and highlight deviations for correction.

SIDEBAR: INVESTIGATING ABNORMALITIES

When investigating abnormalities, consider some of the following questions to obtain a deeper understanding of the cause:

- Did something recently change (man, machine, method, materials, and environment)?
- Are new operators being used?
- Are employees following procedures?
- Is the material within specification?
- Are the equipment, tools and gauges functioning as intended?
- Have there been any recent changes in the process?

Change Management

Change management is about taking a systematic approach to managing process and product changes by understanding the facts behind a change and the impact of the change upon implementation. The purpose is to control the life cycle of all changes occurring within manufacturing operations while ensuring all changes implemented are beneficial and occur with minimal production impact. Changes can help a company become more productive and maintain a competitive edge.

Change is a normal part of business. Unfortunately, in many operations, unapproved and uncontrolled changes are a major contributor to process failures, product defects, and customer dissatisfaction. To avoid this from happening, all changes must be justified, analyzed for their potential impact, and dispositioned as approved or unapproved. If approved, an implementation plan should be executed, and the change benefits confirmed. It's highly recommended that a Change Review Board (CRB) be used to manage complex changes. People selected for the CRB must be knowledgeable of the processes and products under consideration and help prepare prove-out plans that generate the data and information needed to accept or reject a change. When a change is proposed, consider the following cycle of activities:

- **Change request** – understand why a change is being requested. Consider the objectives or expectations of a change on safety, quality, productivity, and cost as well as its potential impact to product (e.g. form, fit, or function) and process performance. *Is the change request justified?*
- **Change Impact Assessment** – complex changes may require sufficient data and information to be properly dispositioned. This can typically be achieved through the development and execution of a change "prove-out" plan used to gather data and information needed to assess the impact of a change on the targeted process or product. Consider the who, what, when, where, and why aspects of change and how much the change will impact the operational status quo. Also consider the potential residual impacts of a change on other aspects of the process, product, and operating system.
- **Change Disposition** – once the requested data and information are available from prove-out plan execution, evaluate the merits of the change request. Accept or reject the change. Communicate the decision, with corresponding justification to all key stakeholders. Clarify the next step for approved changes.
- **Change Implementation** – develop an implementation plan when a complex change has been approved. Consider the impact of change on people, equipment, methods, and materials that will be affected. In particular, if people don't accept or buy into the change, it's not likely to be sustainable upon implementation. For long-duration change deployment, consider periodic governance reviews to assess progress and discuss obstacles to deployment. Be prepared to monitor and adjust deployment plans based on knowledge gained. Don't overlook the need for change acceptance and training.
- **Change confirmation** – when a change is completed, it should be followed up to confirm it was properly implemented, effective, and sustainable. Make sure any changes to manufacturing system operations have been integrated into work standards and procedures.

Change Management Tips

- Have a clearly defined change control process in place.
- Prepare people to work in an environment of change.
- Be transparent when managing change, communicate activities, and encourage employee feedback.
- Define clear ownership for each approved change for implementation.
- For complex changes, develop deployment plans with tasks, ownership, and timelines.
- Periodically confirm the change process is being followed properly.
- Deploy a CRB approach for significant operational changes.
- Target knowledgeable and experienced people for the CRB.
- Prepare a detailed prove-out plan to review and arrange complex changes.
- Schedule periodic meetings to review pending changes for disposition and the status of change management activities.
- Define a change management process that can accommodate simple as well as major changes.

Structured Problem-Solving

Structured problem-solving (SPS) is the application of knowledge, skills, tools, and techniques to solve problems in an organized and disciplined way. It's used to determine root cause of an issue quickly and efficiently so that a corrective action can be implemented to prevent problem reoccurrence. SPS is used in daily shop floor management because it's the most cost effective and efficient approach to problem solving. There are many methods, tools and techniques that can be considered within the problem-solving domain (see Figure 3.7.1).

The method selected for problem-solving is not as important as using reliable data and information for problem analysis and decision making. However, there are several basic steps to SPS that apply to almost all methodologies. They include understanding and clearly defining the problem at the start, analyzing for root cause, implementing an effective countermeasure and following up to confirm effectiveness

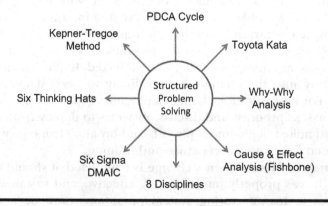

Figure 3.7.1 Problem-solving methods.

and sustainability of the solution implemented. When pursuing problem-solving, it's important to build a strong team that includes one or more subject matter experts who know and understand the process, product, or equipment under consideration.

Structured Problem-Solving Tips

- Always define the problem to be solved. Obtain group consensus for the problem statement.
- Use a structured and proven methodology.
- Consider mapping the process first to obtain understanding.
- Evaluate options before selecting the best solution.
- Don't jump to conclusions, let the data and facts lead you to root cause.
- Decompose a complex problem into small components to solve.
- Consider different perspectives when solving problems.

Just-in-Time Inventory

Just-in-time (JIT) is a manufacturing strategy focused on minimizing material inventory by producing to customer demand. The lean concept of JIT has been challenged over the past several years due to the impact of the Covid-19 pandemic. The pandemic has demonstrated the fragility of JIT when the supply chain foundation, upon which it's built, becomes disrupted. A JIT approach is designed to minimize inventory by ensuring raw materials and finished goods are delivered on time and as needed to maintain a steady production flow. JIT also supports continuous process optimization by eliminating non-value-added activities that contribute to waste and operational inefficiency.

When goods are received as needed, inventories and corresponding costs can be reduced, improving productivity and profit. However, lower inventories decrease material buffers needed to hedge against unexpected problems that can threaten process flow and timely customer delivery. The application of JIT must be complemented by a more accurate production demand forecast that ensures the right material and quantity are available when production needs it. Unfortunately, this will likely put more stress on the supply base and internal workforce since more rigid scheduling reduces manufacturing flexibility.

Policy Deployment

Strategy deployment plays a significant role in organizations with a vision and the desire to realize it. Successful policy deployment requires that people are aware of the company vision and are given clear direction on how to achieve it. This requires a strategic plan, communicated to the entire workforce, that aligns organizational personnel with a set of common goals and encourages them to work in cooperation to satisfy stated expectations. This is done through an iterative feedback loop which leads to goal commitment where the strategic plan is synchronized between all levels

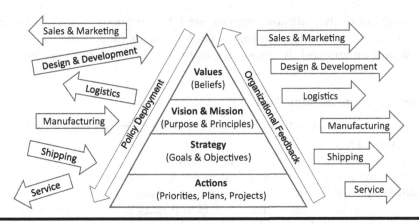

Figure 3.7.2   Policy deployment.

of the organization so that every employee knows the importance of what they do and how their contribution impacts the company's success. See Figure 3.7.2.

A robust strategy is based on a vision, rooted in stakeholder expectations, market needs, and business core competencies. It considers business strengths and weaknesses in addition to customer, supplier, and employee input. Effective policy deployment breaks down the vision into specific areas that articulate the main tactical topics and drivers, so they are understood by all employees and key business partners. The gaps between the current and desired state are identified, analyzed, and actions are prioritized based on business impact. Strategy deployment objectives are then established, based on these priorities, and communicated to the workforce to close existing gaps through targeted projects, company practices, key processes, and procedures. Key process indicators (or metrics) are then used to monitor progress in achieving the common goals and objectives. Employees are trained, mentored, and coached, accordingly.

Policy deployment is important to shop floor management, because it helps employees align their daily work activities with the organizational vision and mission. An aligned workforce is more efficient, motivated, and confident that what they are doing is adding value and contributing to organizational sustainability and goal advancement. This approach also fosters open feedback and employee engagement within the manufacturing system. A tactical implementation plan (TIP) can then be used to help an organization achieve their objectives by providing a shared understanding, action plan, and measures of success. The TIP makes clear what must be achieved, when, and by whom. It serves as a communication platform that offers a set of measurable and time-based goals which can serve as the basis for clear structure, direction, and progress reporting.

Policy Deployment Tips

- Ensure all employees understand the organizational vision, mission, and strategy.
- Ensure employee daily work routines align with organizational goals and objectives.

- Ensure employees understand why they are doing what they are doing.
- Ensure continuous improvement activities focus on strategic project achievement.
- Ensure departmental objectives align with the organization's primary objectives.
- Ensure key process indicators (KPIs) monitor and help drive strategic corporate objectives.

SIDEBAR: ORGANIZATIONAL POLICY

A policy is a set of principles used to guide decisions and achieve desired outcomes. It can be a statement of intent, implemented through a project, procedure, or system. Policies are often defined or articulated through a company strategy and deployed throughout an organization to ensure alignment of all functions and individuals. They can also be a valuable reference when making critical decisions. Business policies are organizational guidelines used to govern employee actions. They can define any number of practices, from the boundaries within which decisions are made to the dress code of company employees. Business policies can also deal with resource acquisitions necessary for the achievement of organizational goals.

Business policies define the rules within which individuals are free to act in pursuit of short- and long-term organizational goals. To be effective, business policies must be simple, clear, concise, and unambiguous. They should align with the vision and mission of an enterprise, be comprehensive enough to provide flexibility in application, promote stability, and inspire confidence. Business policies help shape an organization's mindset, activities, and behaviors reflected in routine business practices. Policies strive to inject discipline and efficiency within operations. This contrasts with a business strategy which is often disruptive and exploratory in nature, driving organizations to act outside the norm to develop new capabilities in preparation for future challenges.

Single-Minute Exchange of Die

When changing the manufacturing line from one product to the next or from one product variant to another, it typically takes time away from production, as materials, fixtures, tooling, and software changes are made, among other line adjustments, to accommodate the next product run. This impacts manufacturing efficiency by reducing the number of parts made per hour during line downtime for the exchange. To minimize this impact, the concept of single-minute exchange of die or SMED was created to bring structure to the process of reducing product changeover times.

The primary objective of SMED is to reduce production line changeover (or setup) time to the shortest possible time without compromising product quality. Changeover time is measured from completion of the last good part of the current variant to the first good part of the next variant produced. The SMED methodology is broken into two fundamental phases, external and internal. The external phase involves activities that can

Line Changeover - Time Reduction Process

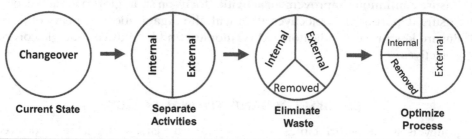

Figure 3.7.3 Changeover time reduction process.

be completed while the process and equipment are still producing parts (no down-time). Internal activities are those that occur while the production line and equipment are down and not producing parts.

SMED workshops are a good way to drive a reduction in changeover time, since they are intended to bring the right people together to focus on eliminating waste with external and internal changeover activities without compromising quality. A good workshop will produce a lean procedure with a sequential checklist of activities to help complete a changeover in the shortest possible time. See Figure 3.7.3 and tips for completing a SMED workshop. More detail on completing an SMED workshop can be found in Appendix 5.

SMED Workshop Tips

- Follow the structured methodology for conducting an SMED workshop.
- Observe one or more changeovers before getting started. Baseline changeover cycle time.
- Document the changeover process with video and pictures.
- Review process steps and workflow (Process Mapping, Spaghetti Diagram, Gantt Chart).
- Consider all possible improvement (waste reduction) measures.
- Look to convert Internal activities to External activities.
- Strive to optimize Internal activities that could not be converted to External activities.
- Optimize all External activities that could not be eliminated as waste.
- Standardize the new "lean" changeover process.
- Periodically audit a changeover process for conformance to the documented standard.

There are additional methods that are discussed in other areas of this chapter including Jidoka, TPM, and layered process audits (LPAs). These are only a few of the many methods used for shop floor management and improvement. It's time to turn our attention to agile manufacturing which can be a difficult yet rewarding concept to apply in the dynamic world of volume production.

Agile Manufacturing

Being agile is about having the ability to move quickly and easily. Agile manufacturing reflects these characteristics within the operating system and can be found in the flexibility of people to do different tasks, to quickly change from making one product type to another, to manage material variations, and the ability to seamlessly adapt technology into the work environment. The efficiency and effectiveness at which a manufacturing team adjusts and handles these types of situations reflect their agility.

Agility is a mindset and skillset that work together to maintain process control within a dynamic working environment. It's about preparing for and responding to the unexpected events that influence daily work conditions. Like many other meaningful things in manufacturing and in life, agile operations require a conscious and intentional effort. It can't be turned on like a light bulb. It becomes part of a company's DNA through strategic planning, execution, and refinement based on experiences acquired through experimentation and lessons learned. It requires people to think and act in ways that simplify processes, continuously striving to making them easier and quicker to execute, monitor, control, and improve. Consider some of the following ways to achieve a more agile shop floor experience:

- Cross-train employees to do multiple tasks.
- Teach everyone to solve problems on their own and in teams.
- Make decisions at the lowest possible level of the organization.
- Solicit ideas from employees on how to do things quicker, easier, cheaper, and better. Vet each idea and deploy what makes sense.
- Design equipment for quick changeovers to run different variants, frequently.
- Design production lines for multiple products using a modular approach.
- Standardize equipment throughout the facility.
- Design products and processes with common/interchangeable parts.
- Reduce the number of components and suppliers. Build strong relationships with remaining suppliers.
- Integrate human intelligence into equipment (e.g. artificial intelligence).
- Install flexible utility supply lines (e.g. compressed air, vacuum, electricity) for easy line layout and adjustments.
- Strive for common packaging.

There are many other ways to exercise an agile existence. What's key to agile manufacturing is to have a knowledgeable, engaged, and empowered workforce with a clear vision and mission to make decisions and take actions within the framework and boundaries of the policies and procedures espoused by the company. There is more on this topic in Chapter 4.3.

SIDEBAR: NEGATIVE SIDE TO COMPANY GROWTH

In a small company, much of the work performed is transparent. Processes tend to be simple and the go-to person for questions or concerns is clear. As companies grow, they tend to put more processes in place, such as the authority for decision making, which is typically assigned to a few key individuals who may not have first-hand knowledge of the decision under consideration. This has the potential for slowing the decision-making process and can result in bottlenecks as busy people may not respond to a request for approval in a timely manner.

As growth continues, more procedures are developed, processes are detailed, and training is scheduled to explain what's needed to do the work for which employees are hired. Sometimes manual processes are turned into electronic ones to make it "easier" to complete. I have observed situations where additional processes were added to correct a problem that was never fully understood, further increasing process complexity and allowing problems to continue to the point people get frustrated and start implementing workarounds. Process is a funny thing; it needs to be properly managed and kept as simple as possible so it can be understood, followed, and effective.

There are also those who like to "hide" behind processes. This is when an individual insists that the process must be followed, as written, so they don't have to do any additional work or make decisions when the current process does not fit or address the existing circumstances. This is when a process starts to fail the company, creating inefficiencies that impact productivity.

Summary

As you can see from the list, some methods have been around for a long time such as PDCA, while other methods are not as clearly defined but are still essential for shop floor management. These methods include deviation management, change management, and Gemba walks which must be tailored for the unique working environment. Often these methods are taught but are best learned through on-the-job training to become comfortable and competent at their deployment. It's good practice to have several experts available to coach and mentor others on the essential methods required to maintain system integrity.

The way manufacturing methods are defined and executed can greatly contribute to the efficiency and effectiveness of the shop floor management system. People who need to know what to do, and when to do it, can find solace and direction in the methods recognized and practiced by an organization. The key is to continuously evolve company methods and procedures, making adjustments based on the changing dynamics of the organization and industry through experimentation, identification of good practices and lessons learned.

Methods can be used to help manage complex processes often by breaking them down into simpler, more manageable components. Knowing when and how to use certain methods is key to good shop floor management. Regardless, it's important to recognize which methods are essential for operational excellence and focus on ensuring those methods are well understood and continuously practiced within the manufacturing system. The responsibility for the application of specific methods can be assigned to individuals or functions via standard work routines and effectiveness confirmed through layered process audits or similar means.

Unfortunately, methods have their limitations, especially if they are not properly deployed. As discussed, methods may require periodic updating and workforce retraining based on observations and lessons learned. Methods should not stifle creative thinking or the desire to experiment to find a better way of doing something.

Key Points

■ Let your vision, mission, and strategy define your methods.
■ Methods serve to define the "how" of doing something.
■ Methods reinforce the desired behaviors and practices of the organization.
■ Methods bring structure and discipline to shop floor operations.
■ Methods set a standard for how certain activities are to be performed.
■ Done properly, methods can help optimize manufacturing performance.
■ Use knowledge and lessons learned to refine your methods.

3.8

Manufacturing Environment

When a flower doesn't bloom, you fix the environment in which it grows, not the flower.

~ Alexander Den Heijer

Objective: Create a comfortable, safe, and flexible work environment.

Overview

A clean and properly controlled shop floor environment is essential for process stability, continued capability, and sustainable improvements. Materials, equipment, and people may be sensitive to temperature, humidity, and cleanliness conditions that ultimately effect productivity and output performance. Factors affecting a work environment can also include building design and age, workplace layout, workstation setup, furniture, equipment design, space, ventilation, lighting, noise, odor, vibration, radiation, and air quality. Many of these factors are discussed in various chapters throughout this book. What's important is not to overlook these potentially subtle impacts on operating performance.

Critical environmental factors affecting product quality and consistency need to be understood and controlled. Standard work routines and layered process audits can be used to highlight and monitor these factors to ensure they remain within specified control limits and deviation management activities can be employed to address significant variations from operating parameters before abnormal conditions lead to process degradation or product defects. Let's consider some of these factors in brief.

DOI: 10.4324/b23307-21

Building Layout/Design

The efficient flow of material through the facility is often dictated by the facility layout, warehouse location, and production line arrangements. For example, having incoming material stored close to the high volume production lines will minimize the movement of material throughout the facility. The effective use of space, including the accessibility to production material drop-off locations and equipment access panels for maintenance, all play a role in creating a flexible and comfortable working environment. Changes to layout and its impact on material flow should be carefully considered prior to execution.

Ergonomics

The disciplined application of workplace orderliness and ergonomic practices can help ensure employee movements are minimized to prevent repetitive motion and workplace injuries. This can also be complemented by standing mats, table height adjustments, and chair comfort. More on the topics of ergonomics can be found in Chapter 2.6.

Safety

There are no "acceptable" accidents in the workplace. An accident is not the result of bad luck, it's a failure of the plant's safety system and deployment of good work practices. Each accident is paid in blood by employees. Therefore, we must take advantage of lessons learned from each accident, giving it the attention necessary to avoid reoccurrence. Steps to prevent reoccurrence should include on-going safety training, integration of lessons learned into daily work practices, the application of safe work procedures, and exercising good judgment. The safety mindset must be reinforced by management's commitment to promoting accident prevention in the workplace. See Figure 3.8.1.

Traditional management of safety practices in the workplace focused on lagging indicators such as accident rate and reacting to events to prevent their reoccurrence. A more proactive, preventive, and predictive approach to safety, reflecting manufacturing excellence, should focus on leading indicators such as continuously promoting a safety culture, gauging the effectiveness of existing safety systems and practices, taking action to prevent near-misses from becoming future accidents, and the continuous implementation of longer-term safety improvements. The more you care about safety, the less likely accidents will occur. Some tips for promoting a safe workplace and culture include:

- Talking about safety performance in team meetings.
- Prioritizing safety improvement actions.
- Including safety in Gemba walks.
- Using management as a role model for wearing personal protective equipment (PPE).

Figure 3.8.1   Safe workplace.

- Highlighting equipment safeguards and their proper use.
- Periodically confirming the required use of PPE on the shop floor.
- Displaying safety key process indicators (KPIs) on shop floor information boards.

Lighting

Proper lighting can improve employee vision, alertness, avoid disruptive shadowing and improve color recognition during product assembly and inspection. Consider the availability, intensity, and positioning of lighting at shop floor work areas and inspection stations. Type of lighting, lighting fixtures, wall and ceiling color, as well as lighting height above employees, can all play a part in sufficient area illumination. Also consider sensored and high-efficiency lighting (such as LED), for energy conservation and economic purposes. Note that the Occupational Safety and Health Administration (OSHA) oversees and regulates workplace safety standards, which includes establishing minimum industrial lighting standards for manufacturing settings.

Temperature

Consider the facility's temperature requirements relative to the optimal use of production machines, product stability, and measurement equipment. A comfortable ambient temperature will minimize employee discomfort in being too hot or cold while working. Comfortable employees are generally more productive. In operations, where equipment overheating or product spoilage (e.g. food and medicines) is a concern, sensors can be installed to warn of a significant temperature deviation. Proper temperature control can be important for maintaining equipment lifespan, measurement accuracy in a laboratory, and for product stability, especially where chemical reactions may occur.

Ventilation

This is an important activity for providing a healthy and safe workplace for employees. Ventilation systems are designed to remove air contaminants from the building environment and replenish it with fresh outside air. This can be critical in certain facilities where flammable vapors, welding fumes, dust, mold, oil mists, and toxic chemicals are present. In specific facilities, contaminant removal is necessary to reduce the likelihood of fire and explosion hazards. Contamination control is also important to avoid employee discomfort and illness. Ventilation equipment maintenance and monitoring of air quality are recommended to ensure the ventilation system continues to work properly, without fail.

Air Quality

Indoor facility air quality is important for employee comfort and health since indoor pollutants can pose an environmental risk to occupants, if not properly managed. Under certain conditions, workers are required to wear respirators for protection against excessive dust, fog, smoke, gases, chemical vapors, sprays and other concerns, when air exceeds OSHA's permissible exposure limit (PEL) for airborne contaminants.

If excessive amounts of dust are evident on the shop floor and flat surfaces, or damp walls are apparent from airborne mist particles, you may have a contamination problem. This may also hold true when musty or burning smells are detected while breathing. In these cases, consider air quality testing to understand what actions may be required to correct a potential problem. On-going air quality monitoring is recommended to ensure employee comfort, product quality, and regulatory compliance.

Air Current

The existence of air currents in the facility, especially for labs and cleanrooms, may need to be controlled. The accuracy of measuring devices, such as those in a metrology lab, may be affected by air flow from an open door. Air pressure in a clean room may also need to be regulated to maintain a cleanliness classification standard.

Noise

Noise is considered hazardous when it reaches or exceeds 85 decibels. According to an Occupational Hearing Loss Surveillance (OHL) by The National Institute for Occupational Safety and Health (NIOSH)[1], about 46% of all workers in manufacturing have been exposed to hazardous noise and 28% of noise-exposed manufacturing workers report not wearing hearing protection. Industrial noise is typically caused by machinery, construction, and vehicles. Prolonged exposure to noise can cause stress, fatigue, and productivity loss, not to mention communication issues.

Noise can be managed through building design using noise-absorbent materials, investing in quiet equipment, and machine maintenance to avoid their noisy deterioration. Make sure OSHA standards are being met and noise protection equipment is available for those who need it, or want it, for comfort. I have encountered workers who were unable to work in a particular area of a plant due to constant noise, regardless of the PPE available to mitigate its impact.

Humidity

Different manufacturing facilities experience different humidity-related issues depending on the specific climate, equipment, and products employed. Excess humidity can result in condensation, mist, or mold, which can lead to product and equipment damage or deterioration. Condensation on facility floors can become a slip hazard. Too little humidity can lead to employee discomfort, degradation of machine performance, and impact material dry times. There are many other examples of humidity's effect on manufacturing operations. Regardless, action can be taken to control facility humidity, in dry and wet climates, which may impact material storage, equipment stability, and employee comfort. When falling outside of a comfortable or specified humidity range, typically between 50% and 60% for humans, humidifiers can be used to add moisture and dehumidifiers employed to extract it, as needed.

Facility Space

Do employees have sufficient space within which to work, take breaks, store their personal items, and change into work attire? Is there adequate space for material storage and movement, training facilities, prototype labs, and meeting rooms? Is access to the outdoors available to employees that want to take a break outside or smoke? These and many other considerations need to be taken into account when creating a facility layout or repurposing floor space. The key is to strike the right balance between employee comfort and operating needs.

CASE STUDY: THE VALUE OF SPACE

I once worked in a small manufacturing facility that was filled to production capacity. It was wonderful to see nearly all the available manufacturing space occupied with operating equipment, but it also created a dilemma; there was no "green space" to grow the business. Excess equipment and packing boxes consumed valuable warehouse space and clutter was accumulating on the shop floor as people searched for places to store less frequently used production tooling and fixtures. The only course of action, at the time, was monthly rental of cargo trailers for external storage of selected items. Although space costs money, there must be a balance between space needed to work, for employees, storage, and organizational growth. In short, be protective of "free" space, as it becomes more valuable the less of it you have.

Vibration

Equipment vibration can cause worker discomfort, premature equipment wear, and damage. Vibrating equipment can also create excessive noise, consume more energy, and compromise safety. Equipment vibration can sometimes be an early warning sign of equipment degradation which can trigger remedial action to prevent unexpected equipment failure. Consider opportunities in the plant to address abnormal levels of equipment vibration, especially in the measurement labs where equipment sensitivity to the environment may impact measurement accuracy and precision.

There are many other environmental factors to consider in a plant setting that stem from nature and man-made activities (e.g., radiation, radon, power lines, chemical vapors, etc.). The key is to be aware of these factors, understand their impact on people, equipment, product quality, and process performance and take steps to control them. When understood and properly managed, they are likely to have minimal impact on employees and operating performance.

Key Points

- Environmental factors play a key role in employee comfort and productivity.
- Your five senses of sight, smell, touch, taste, and hearing can be used to assess the normalcy and adequacy of the manufacturing environment.
- A properly-maintained and well-functioning facility requires resources and attention.
- People forget the important of a facility's contribution to productivity when all is running well. This is, as it should be!

Source

[1] NIOSH:https://www.cdc.gov/niosh/topics/ohl/manufacturing.html#:~:text=About%2011%25 %20of%20all%20Manufacturing%20workers%20have%20tinnitus.&text=About%2020%25% 20of%20noise%2Dexposed,have%20a%20material%20hearing%20impairment.&text= Hearing%20impairment%20is%20hearing%20loss,hearing%20impairment%20in%20both% 20ears

SHOP FLOOR IMPROVEMENTS

Complacency is the forerunner of mediocrity. You can never work too hard on attitudes, effort and technique.

~Don Meyer

Objective: Eliminate process waste while improving material and information flow.

Overview

A business can't survive long-term without sustainable efficiency improvements to combat the constant threat of global competition from high- and low-cost manufacturing countries. This reality can't be ignored. Core to any competitive manufacturing practice is the concept of continuous improvement. Part 4 covers various methods, workshops, and activities to pursue efficiency improvements with process optimization and operational excellence in mind. Regardless of the methods discussed, continuous improvement should be a management-driven activity requiring daily engagement. Employees, at all levels of the organization, must be empowered to seek and share their improvement ideas, in a welcoming environment that embraces each idea while recognizing and rewarding employee contributions.

Shop floor improvement encompasses the elimination of non-value-added activities (e.g. waste) and removal of obstacles to process flow. There are several approaches to driving productivity improvements once process stability, capability, and control are demonstrated. They include:

- **Incremental improvements** – Small (and sometimes large) continuous changes that happen every day to make work more efficient. These changes are typically in response to an observed opportunity for improvement or employee idea. They tend to be cheap, easy, and quick to implement.

DOI: 10.4324/b23307-22

- **Kaizen events** – These events are structured and involve a specific, targeted approach to continuous improvement that may need several hours or days to address a concern or pursue a promising opportunity. A Kaizen event typically occurs in response to a known concern, leading to one or more sustainable process improvements.
- **Jishuken workshops** – These workshops are usually in response to a desire to elevate operational performance within a production line or work area. Each workshop is a "self-study" approach to improvement, where multiple days are dedicated to observing, collecting, and analyzing data and information to understand the current process state and highlight opportunities for efficiency improvement. Jishuken workshops actively "look for" problems and opportunities, making it a proactive approach to productivity improvement.
- **Agile methods** – Disruptive technologies are rapidly changing the business climate with mature organizations becoming increasingly marred by complexity and slow innovation. Competition is coming from smaller, more agile organizations able to rapidly deliver products that meet customer needs. The speed of change is forcing larger organizations to adopt a more agile approach to hedge against competitors looking to take their market share. Adopting an agile approach to business helps organizations manage challenges posed by constant change, market uncertainty, and increasing complexity. In practice, agile leverages customer experience, through fast and transparent feedback loops, to quickly provide iterative deliverables in response to a disruptive and demanding environment driven by rapid advances in technology.

Part 4 was written to provide different types of methods, activities, and workshops that can help manufacturing facilities achieve and maintain process stability while continuously striving to improve their operational performance. It's intended to promote an action-orientated approach to continuous efficiency improvement through methods designed to enhance employee understanding, develop their lean skills, and augment the capability of an organization to drive shop floor efficiency improvements for competitive advantage.

To this end, Part 4 has been divided into five chapters. Chapter 4.1 is focused on Product and Process Development. This is important for shop floor management because many of the decisions made and actions taken during the product and process phases of development affect daily shop floor activities during production. Chapter 4.2 is orientated around some of the more common methods, tools, and techniques used to drive incremental and sustainable productivity improvements such as process mapping, workplace design, Kaizen events, and Jishuken workshops. Chapter 4.3 shifts our attention to agile methods where we explore the practices, lifecycles, and techniques associated with a flexible and highly adaptive mindset. This chapter is followed by Chapter 4.4 which considers ways to streamline process execution by touching on the fundamental practices that nudge manufacturing activities along the path toward optimization. The final Chapter 4.5 brings many of the elements discussed in the previous chapters together to provide more context and structure for the pursuit of manufacturing excellence. The objective is to augment the reader's understanding of the information provided and enhance their ability to apply what has been presented in resourceful and practical ways. Let's get started!

4.1

Product and Process Development

> Quality planning consists of developing the products and processes required to meet customer's needs.
>
> ~ Joseph M. Juran

Objective: Develop robust products and processes that steer lean and agile manufacturing operations along the path to optimization and excellence.

Overview

The quality of product and process development activities will have a significant impact on shop floor management. As a result, it is important that manufacturing is involved early in the development process since production system stability and capability depend on it. The ability of manufacturing to satisfy engineering tolerances and customer specifications is based on the integrity of the development process rooted in the design team's knowledge, experience, and discipline in following a proven product launch procedure.

Methodologies such as design for manufacturing (DfM), design for six sigma (DFSS), and the development of a failure modes and effects analysis (FMEA) help to reinforce a manufacturing optimization and excellence mindset. The primary objective of these product development techniques is to reduce the impact of "noise" factors (sources of variation) that can significantly disrupt a process from delivering a quality product, on time, to customer requirements. A lack of product stability can wreak havoc on the manufacturing floor. Therefore, it is important that the manufacturing team is well aware of what is in the design pipeline and continuously engages with the design team, throughout the development process, to influence the outcome of the final product and process prior to production release. In essence, the voice of manufacturing needs to be heard, early and often, during the development process.

DOI: 10.4324/b23307-23

Once a product is released to production, manufacturing must live with what they inherit. Considering this, we will overview several methods that require manufacturing team awareness and engagement in the development process to proactively influence the design team's work activities. Let us start by understanding the benefits of DfM in contributing to shop floor management excellence.

Design for Manufacturing

DfM is based on the premise that manufacturing knowledge and experience are considered early in the product development process, to simplify, refine, and optimize product design for lower cost and ease of assembly. DfM involves taking a more structured and cooperative approach between the manufacturing and design teams to identify high-risk issues related to new, unique, and difficult aspects of a design (e.g. NUDs) and working together to mitigate those risks during development. The objective is to study key NUDs in depth to understand their impact on manufacturing capability and take action to optimize the design by making it simpler, and low-cost for production. This deeper knowledge comes from team analysis of key design features, prototype builds, testing, and experimentation, leading to the selection of suppliers, equipment, components, and materials that will achieve the desired engineering tolerances and assembly expectations for lean and agile manufacturing. A deeper understanding of product and process design fundamentals can also help to develop more robust processes with less sensitivity to sources of variation. See Figure 4.1.1 for an overview of the DfM process.

DfM must consider production volumes, product complexity, design tolerances, building materials, and the ability of the design to comply with good

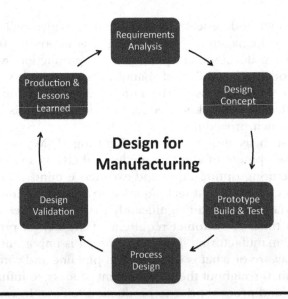

Figure 4.1.1   Design for manufacturing overview.

manufacturing principles and practices. Understanding a product's material, electrical, mechanical, optical, and thermal properties can also play a key role in DfM. Other characteristics like color, flammability, and environmental resistance may need to be studied to ensure compliance with safety, regulatory, and quality standards.

As stated, DfM is about reducing manufacturing costs without compromising quality. Consider minimizing part count to reduce material, engineering, and shipping costs. Use parts and materials that are readily available (e.g. off the shelf) since they are typically cheaper and easier to obtain than customized parts. Consider modular assemblies, minimize part handling, and avoid reorienting parts during machining and assembly operations. Joining materials can also be simplified and less expensive by avoiding the use of screws and fasteners. Finally, investigate reducing the pieces of equipment and manufacturing steps "needed" for product assembly. In certain cases, this can result in significant cost savings.

It is important to remember that part tolerances should be carefully considered since unreasonably tight tolerances can cause unnecessary scrap while excessive tolerances can allow unacceptable parts to reach the customer. Exercising the full tolerances allowable for producing a quality product will help lower tooling costs, reduce defects, increase yields, and improve the ease of manufacturing. To define reasonable manufacturing tolerances, consider the manufacturing process, the material selected, and the product sensitivity to variation. Data analysis of similar products already in production can aid in this regard.

In the end, the primary goal of DfM is a low-cost, high-integrity design that is easy to manufacture and meets customer requirements for the right price. It is a grave error when the design requirements for manufacturing are known but not communicated or are simply ignored by the design team, leaving development outcomes to chance.

Design for Manufacturing Tips

- Prepare manufacturing design guidelines based on engineering experience and lessons learned. Follow the guidelines, making improvements when warranted. All exceptions to the design guidelines should be justified with data and discussed during design reviews.
- Understand customer wants, needs, expectations, and application requirements prior to starting the design process.
- Consider all part requirements and tolerances. Reflect on how these factors will impact manufacturing capability.
- Start design discussions with key stakeholders early in the development process, well before production tooling is ordered.
- Conduct critical design reviews prior to design freeze. Promote open and honest discussions, early in the design process, to elevate and address problems when they are easier and cheaper to manage.
- Build design prototypes as soon as possible, preferably on a production line. A tremendous amount of information can be gained by analyzing a physical object and the process of building it.

- Have a manufacturing engineer assist with sample builds. Their awareness, knowledge, and insight can be invaluable to a development team. Record all things that go right and wrong during the build. Prioritize the things that went wrong for action.
- Make all design changes before the design freeze. It is cheaper and easier to make changes on paper than in tooling or production.
- Look at all aspects of the design (e.g., components, sub-systems, and systems). Understand what is new, unique or difficult about the design. Prioritize your development resources on these topics without forgetting about the basics.
- Understand the functional elements of product and process design, the know-how and know-why.
- Rigorously test the product early and often. Use lessons learned to augment product and process design.
- If available, deconstruct a similar/competitive product for design insights.
- Have a process design problem? Talk to manufacturing, they may already have a solution.

Design for Six Sigma

DFSS is used to increase product or process robustness and reliability, at the point of the design. The goal is to predict product/process behavior through modeling and simulation, looking for opportunities to reduce sensitivity to sources of variation in the production environment. A more robust product and process helps reduce the obstacles to achieving manufacturing stability and maintaining process control. An overview of the DFSS methodology is provided in Figure 4.1.2.

To maximize the benefits of this methodology, it is best applied to new, unique, and difficult (NUD) aspects of a design, early in the development process. This activity helps project teams understand and evaluate high-risk elements of new designs

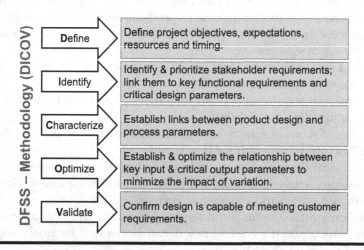

Figure 4.1.2 DFSS methodology (DICOV).

and allows them to act early, when it is easy and cheap to mitigate potential risks before they become chronic manufacturing problems. For example, part specifications can be optimized prior to design freeze to ensure the right tolerances are being defined based on early simulation and modeling results. If DFSS is executed properly, design problems can be highlighted early and prevented from disrupting the manufacturing process. In addition, the right balance between product cost and functionality can be achieved with early feedback from prototyping and testing within the targeted application environment.

Design and Process FMEA

The FMEA is a process analysis and risk assessment tool. It prioritizes potential failure modes based on their severity (S), frequency of occurrence (F), and detectability (D). Failure modes with a high "risk priority number" or RPN, determined by event severity, frequency, and detectability (RPN = SxFxD) will be targeted for mitigation and possible elimination. Essentially, the FMEA is intended to identify design and process weaknesses for action to reduce the likelihood of defect occurrence and, if they do occur, preventing defects from escaping undetected to the customer. It can also serve as a tool for documenting current knowledge and tracking actions for continuous design and process improvement. When done right, an FMEA will improve product and process quality, reliability, and safety while avoiding excessive time and cost associated with managing failures. See Figure 4.1.3.

In summary, proactive methods such as DFSS, DfM, and FMEA help both development and manufacturing teams identify and mitigate product and process design weaknesses before they become chronic manufacturing problems that can negatively impact shop floor activities.

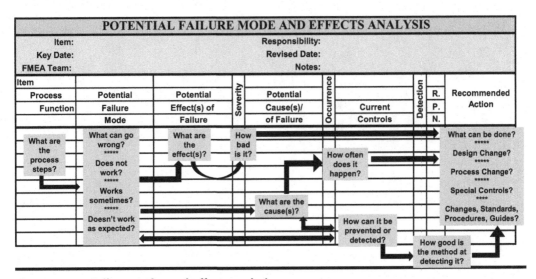

Figure 4.1.3   Failure modes and effects analysis.

Key Points

- Ensure early involvement of manufacturing engineers in the design and development process.
- Apply lessons learned from past development efforts to current and future design activities.
- Identify high-risk design elements during the various stages of development for study and mitigation.
- Understand the interactions between product design and process development to identify opportunities for improving manufacturing productivity.
- DfM should consider design, process, material, environment, testing and compliance requirements.
- An FMEA is an analytical methodology used to ensure potential problems have been considered and addressed throughout product and process development.
- The Design FMEAs should be initiated early in the design process with the Process FMEA started before tool and equipment development.
- An FMEA should be considered for a new design, new technology, new process, and significant modifications to an existing design or process. This includes changes to an existing design or process targeted for a new application environment, manufacturing location, or usage profile.

Source

https://news.ewmfg.com/blog/manufacturing/dfm-design-for-manufacturing

Productivity Improvement

Productivity is never an accident, it is always the result of a commitment to excellence, intelligent planning, and focused effort.

~ Paul J. Meyer

***Objective*:** Produce more output with the same or less input.

Overview

Productivity improvement is getting more product output with the same or less input. It is a measure of output per unit of input. A holistic view of productivity improvement requires a long-, medium-, and short-term approach to manufacturing management. The long-term approach centers on improvements made to product and process design and execution. This can include building a new plant, manufacturing equipment design, and selection of machines, tools, and gauges for application in production. A medium-term approach to operational efficiency is more focused on product simplification and reduction of manufacturing variety. Short-term improvements can be realized through the leaning of processes and procedures, better logistics planning, and the efficient management of labor. These are only a few ways to view productivity improvement.

Companies must engage in continuous improvement as part of their daily work routines. Simply maintaining existing operations is not an adequate or sustainable

DOI: 10.4324/b23307-24

business strategy, especially when competing in a global market. To remain relevant, companies must exercise a process control and continuous improvement mentality as an integral part of their DNA. Continuous incremental improvement includes the elimination of non-value-added activities (e.g. waste) from processes and the removal of obstacles to material and information flow. There are many approaches to driving improvements. We will discuss some of them in the following sections.

Incremental Improvement

Incremental improvement is the process of making many changes to enhance manufacturing performance and productivity. Many of these changes are low-cost and low-risk changes, suggested and implemented by employees, to move the organization closer to a state of optimization and excellence. Incremental improvement can come from employee ideas, process audits, observation, Kaizen events, Jishuken workshops and many other avenues. These changes may be incremental but they are necessary, on-going and must be sustainable.

The plan-do-check-act (PDCA) cycle is a common and practical way to facilitate an incremental approach to manufacturing improvement. It is a simple concept that can be applied to most improvement opportunities by planning, implementing (do), evaluating (check), and taking necessary action (act) to ensure improvements have been realized properly and are effective. Improvements are not just about changing processes and procedures, they are as much about obtaining the buy-in and commitment of people to accept, implement, and sustain all approved changes by making them an inherent part of their daily work routines. By making the PDCA cycle a known and frequent practice of daily shop management, you set the stage for creating a mindset of continuous, never-ending improvement.

Sustainable Improvement

Sustainable improvements require a stable operating foundation upon which changes can be implemented and take root. Stability starts with the implementation of effective standard procedures and is maintained through process control. Process control requires structure, discipline, and accountability. Structure comes in the form of processes, procedures, and methods. Discipline is part of organizational culture and is reflected in employee behavior. Accountability is the act of ensuring employees follow the systems developed to maintain operational performance.

Sustainable improvements require planning, execution, monitoring, and control. Opportunities for improvement must be identified, vetted for their value-contribution, and accepted or rejected. If accepted, improvements must be implemented, confirmed effective, and controlled. Sustainable productivity improvements can be identified in some of the following ways:

▪ **People** - do you have the right people, with the right skills, to do the work expected now and in the future? If not, look to hiring new talent or developing existing employees to help move the organization toward a leaner, optimized and excellent state of operation. People bring knowledge, skills, and ideas to the workplace; use them.

■ **Processes and Procedures** – processes often become confusing and convoluted over time when little to no attention is paid to their upkeep, especially when changes are continuously being made. Seek out opportunities to streamline key processes and corresponding procedures. Eliminate irrelevant steps and workarounds to problems that have occurred over time. Remove obstacles to material and information flow. Work to reduce the impact of bottlenecks on process cycle time. Strive to keep processes and procedures simple and understandable.

■ **Equipment** – planned downtime for maintenance usually costs much less than unplanned downtime due to worn or failed equipment. Maintain equipment in good working order. Prepare and respect the equipment maintenance schedule. Take advantage of unexpected line downtime to "catch up" with or to "get ahead" of scheduled maintenance. Store critical spare parts in inventory. Look to improve equipment capability and capacity. Consider replacing old equipment with the latest technology, if the benefits outweigh the cost.

■ **Automation** – automation is a powerful tool for increasing operational efficiency and reducing production errors. New software solutions can help to improve logistics planning, inventory control, and monitoring workflow, among other things. Technological advancements often change the skills needed for certain tasks. Additional education and training may be required for workers to keep up with changes in technology and its application in manufacturing.

■ **Education and Training** – employee development is good for employee retention, especially for new employees who may need time to become comfortable and proficient at assigned tasks. Training sessions for key operators may be required when procedures are updated, or new equipment is installed. Do not limit training and education to certain employees. Operations will run more smoothly if everyone understands workplace policies, procedures, and practices that drive daily work activities. Offer formal educational opportunities to employees who wish to advance their educational degrees, knowledge, or skills. Productivity improves when people know how to do their job properly.

■ **Workspace Organization** – create the optimal placement for tools and equipment needed for the job. Remove unneeded or unused tools and equipment (e.g. clutter) from the workplace. Create organized storage to reduce the time needed to find materials, documents, and equipment. Layout the manufacturing floor to maximize operational efficiency. Reduce operator and material travel distances. Excess people and material movement is a sign of poor organization and can significantly impact product cost and lead time. Consider techniques such as 5S and Kanban to reduce production delays and increase efficiency.

■ **Material Inventory** – if you have too much inventory, it is probably costing you money and consuming limited space to store it. If you are short of material, you will likely delay or stop production as you wait for more material to arrive.

Software can help track inventory and create automatic notifications of pending shortages. Certain applications will allow material suppliers' direct access to your inventory counts which can trigger them to deliver needed supplies in a timely manner. Proactive manufacturing operations make predictions about the impact of shortages and put plans in place to mitigate potential delays.

■ **Supplier Relationships** – keep track of raw material rejection rates, declining component quality, and late deliveries so you can immediately address these issues with suppliers. It may be necessary to find new suppliers when chronic problems persist. If you know one of your vendors is undergoing a significant change such as disaster recovery or being sold to another company, request a plan that ensures your company's deliveries will continue to be met as contracted.

SIDEBAR: LEVERAGING ORGANIZATIONAL KNOWLEDGE

Knowledge is often acquired through training, but training alone will not produce great results. Training can provide the how and why, but deep learning is often complemented by meaningful action to demonstrate a proven practice and reinforce its value to achieve desirable outcomes. Training creates awareness and understanding, but knowledge acquired through training is waste, if it is not used to advance the organization's intended objectives. The exploitation of knowledge through the application of the right skills can generate activities that move an organization closer to operational excellence. Thus, skillful application of knowledge can create lasting value.

Developing the knowledge and skills of organizational personnel, to drive efficiency improvements, can be realized through a learning/action model that applies knowledge in real-time and in real-life situations. Motivated people must have the knowledge, skills, resources, and time to implement sustainable change. Therefore, workshops, which combine training and the application of knowledge, in a learning/action model, are an effective approach for developing people and changing behavior when shaping organizational culture.

Improvement Methods

There are many different methods for driving improvements. Engaging in improvements allows a business to reduce costs and remain competitive by continuously enhancing operational efficiency. There are many ways to pursue improvements, some formal and others informal. Less formal, yet structured approaches comprise the application of PDCA, quality circles, and employee idea management. More formal approaches included Kaizen events, Jishuken workshops, Single-Minute Exchange of Die (SMED) workshops, Six Sigma DMAIC (define, measure, analyze, improve, and control), and Toyota Kata projects. These and other methods will be discussed in the following pages. Let us begin the discussion with quality circles since we have covered PDCA in past chapters.

Quality Circles

Quality Circles usually involve a group of individuals or employees who meet periodically to solve work-related problems or issues to improve performance in their area of responsibility. Teams typically contain three to six people who work to solve problems and implement improvements. These teams are led by competent individuals (e.g. facilitators) who are specialists in their field and trained to work with people in problem identification, information gathering, and data analysis (e.g. basic statistics), solution selection, and implementation. Quality circles tend to be more successful when supported by the company's senior management versus a more "grass-roots" approach to deployment. It is common for quality circle teams to use the seven basic tools of quality including cause & effect analysis (e.g. Ishikawa or Fishbone diagrams), pareto charts, histograms, run and control charts, scatter plots and flow charts. Reference Figure 4.2.1.

Employee Idea Management

Employees are one of the best sources for improvement ideas. They understand their jobs and are more likely to see opportunities for improvement. Every organization should encourage the sharing of employee ideas and have an organized system in place to ensure all relevant ideas are considered. Employees should be recognized for their ideas and rewarded for an idea that has been accepted, implemented, and delivered proven results. Ideas should be judged by their impact and benefit to the organization, not by the individual who suggested them. Make sure there is a process in place that captures, dispositions, approves implements, and rewards employees for their contribution to productivity improvement.

Set-Up and Changeover (SMED)

Reducing cycle time is a primary focus of waste elimination. One way to accomplish this is by reducing production line changeover time. Changeover time is the time to change from one product variant to another. This process is broken down into two main activities, internal and external set-up time. Internal set-up time is the time during a changeover when a process or machine must be stopped to allow for transition. External set-up time includes all other changeover activities performed while the machine or process is running such as preparing for or following up with any changeover activities. The objective is to transition as many internal set-up activities to external activities followed by reducing each activity to the shortest possible time. Changeover time reduction allows for more frequent changes in product variants within a given time, facilitating the reduction of lot sizes and minimizing excess inventories. Capacity is consumed by changeover time, thus, reducing the time for changeovers will increase productivity. See Appendix 5 for details on conducting a SMED workshop.

Six Sigma DMAIC

The DMAIC methodology is a disciplined, data-driven approach for improving the quality and efficiency of an organization's operational and transactional processes.

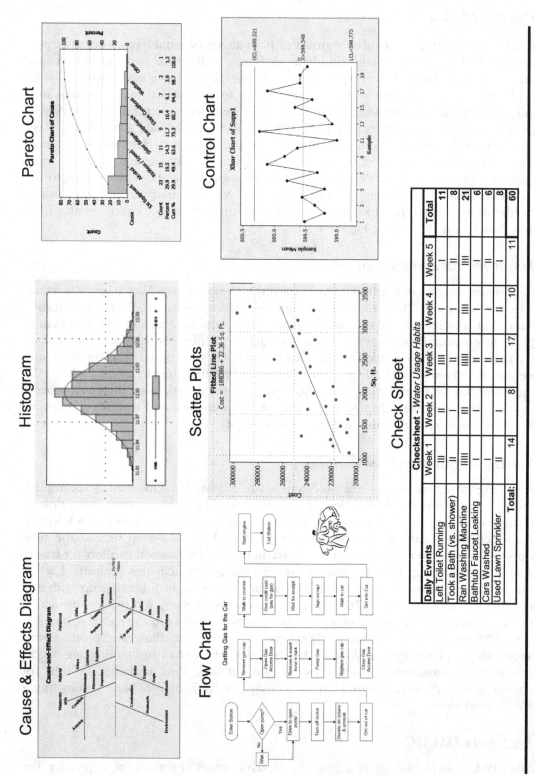

Figure 4.2.1 Seven basic quality tools.

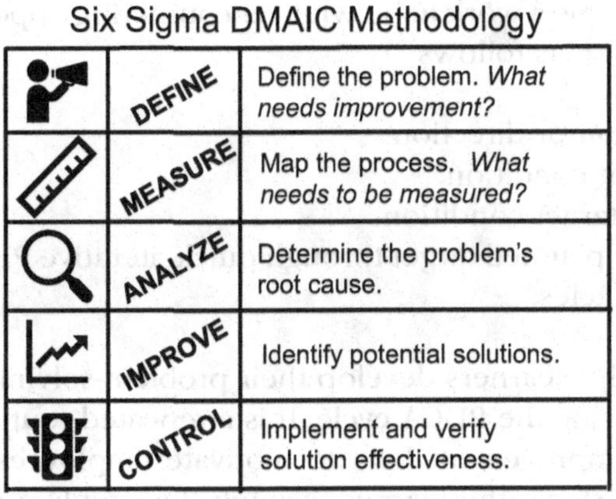

Six Sigma DMAIC Methodology

📢	DEFINE	Define the problem. *What needs improvement?*
📏	MEASURE	Map the process. *What needs to be measured?*
🔍	ANALYZE	Determine the problem's root cause.
📈	IMPROVE	Identify potential solutions.
🚦	CONTROL	Implement and verify solution effectiveness.

Figure 4.2.2   Six Sigma DMAIC methodology.

It is a measurement-based methodology that can be used to significantly reduce operational errors and product defects by decreasing process variation and enhancing process robustness. This is accomplished through the deployment of statistical and analytical tools proven effective for problem-solving and process waste reduction. Some of the key principles behind the Six Sigma methodology include:

- A focus on customer requirements.
- A disciplined, proven methodology for efficiency improvement.
- Understanding the current state of a process targeted for improvement.
- Working to reduce process variation, errors, and defects.
- Data-based decision-making.
- Use of reliable data and factual information.
- Permanent elimination of problems through root cause analysis and corrective action.

DMAIC leads to sustainable process improvements reinforced by the control phase. If this methodology is embraced and applied to core processes, it can create more robust operating systems, improve manufacturing performance, and support the journey to operational excellence. See Figure 4.2.2 for an overview of the Six Sigma DMAIC process.

Toyota Kata Projects

Toyota Kata takes a systematic approach to solving problems while concurrently coaching employees to become better problem-solvers. The Kata improvement methodology was created by Mike Rother [1] after studying the Toyota production system. It's a two-part concept of process improvement (Improvement Kata) and employee development (coaching kata). The improvement kata is a four-step routine. It drives continuous improvement through the fundamental problem-solving method of PDCA and is expected to become a work habit when encountering opportunities for improvement. The four steps are as follows:

- Determine a vision or direction.
- Grasp the current condition.
- Define the next target condition.
- Move toward the planned target through quick, iterative PDCA cycles to uncover and remove obstacles.

The coaching kata helps learners develop their problem-solving acumen by working to remove obstacles using the PDCA cycle. It is a repeated routine where leaders and managers teach the improvement kata to motivate employees in the organization. The teacher or coach steers the learner through the problem discovery and resolution process, without providing solutions, by asking the learner questions before and after each PDCA cycle. See kata-coaching questions in Figure 4.2.3. Learners develop a plan with their coach that is executed, reviewed and followed up with additional action, as needed, until an obstacle is eliminated. This repeated process requires the learner (or improver) to think for themselves in working to overcome obstacles to improvement.

Value Analysis

Most processes contain non-value-added work or waste which consumes unnecessary time and money. Value analysis is a proven method for reducing process cycle time and improving efficiency in manufacturing and service (transactional) operations. It starts with process mapping to visualize the type and sequence of

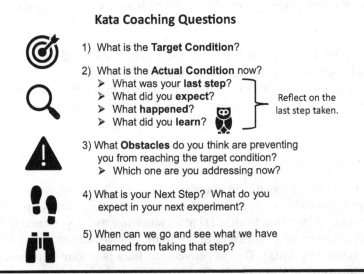

Kata Coaching Questions

1) What is the **Target Condition**?

2) What is the **Actual Condition** now?
   - What was your **last step**?
   - What did you **expect**?
   - What **happened**?
   - What did you **learn**?

   Reflect on the last step taken.

3) What **Obstacles** do you think are preventing you from reaching the target condition?
   - Which one are you addressing now?

4) What is your Next Step? What do you expect in your next experiment?

5) When can we go and see what we have learned from taking that step?

Figure 4.2.3 The five Toyota Kata questions and four reflections.

activities in a process. This visualization is used to classify work activities into value-added, non-value added and business requirements. Non-value-added activities are considered process "waste" and become a target for elimination. In essence, value analysis is used to identify the greatest sources of waste in the manufacturing system.

Value analysis starts with a detailed understanding of the process under analysis, from a customer view of value-added activities. Value-added activities are the transformational changes to a product, service, or process for which the customer is willing to pay, or any task or activity that is considered "valuable" by the customer. The perception of value-add is based on who you define as the customer. On the other hand, business requirements are the essential activities necessary to function as a business such as ordering materials, paying taxes, performing credit checks, and administering employee benefits. Non-value-added work activities contribute to organizational waste. It can be work not completed correctly or activities such as inspections, reviews, approvals, and scrap that do not add any "value" to the products, services, or results being offered. The following sequential steps can be used to complete a value analysis along with a visual of these steps in Figure 4.2.4.

- **Step 1: Process Mapping** – prepare a detailed map of activities within the process being analyzed. Document all activities of a process (e.g. on post-it notes, one activity per post-it). If available, record the time it takes to do each activity. Include wait time as an activity to document the waiting period.
- **Step 2: Process Analysis** – review the process map for obvious problem areas. These areas harbor waste:
  - *Disconnects* – poorly managed handoff points or inadequate communication of requirements by a supplier and/or customer.
  - *Bottlenecks* – process points where volume exceeds capacity, slowing process flow. Bottlenecks can impact timeliness and adequacy of delivery quantities if not properly managed.
  - *Redundancies* – process points where activities are repeated or duplicated (e.g. entering the same data in two different systems).
  - *Rework loops* – process points where significant work volume is fixed, corrected or repaired.
  - *Decisions / Inspections* – process points where choices, evaluations, checks, or appraisals could create potential delays.
- **Step 3: Activity Classification** – Separate process activities into value-added, non-value added, and business requirements.
  - *Value-added* – activities for which customers will pay!
  - *Non-value-added* – activities that are unnecessary, redundant, or inefficient. They can include inspections, reviews, approvals, waiting, and rework, among others.
  - *Business Requirement* – operationally value-added work. Work that keeps the process running but may have no customer value.
- **Step 4: Highlight Waste** – Add up the number of non-value activities (e.g. process waste).
- **Step 5: Categorize Waste** – Separate non-value activities into categories (optional).

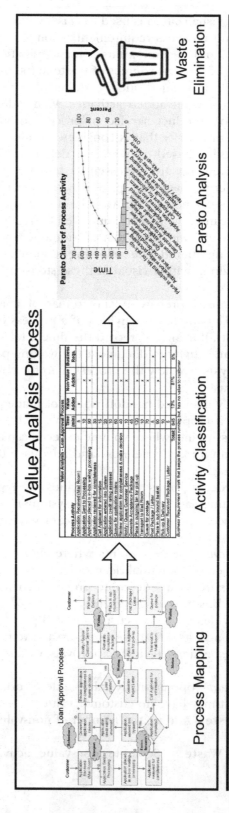

Figure 4.2.4 Value analysis process.

- **Step 6: Prioritize Waste** – Rank (prioritize) non-value activities from most to least significant. A bar graph (or Pareto Chart) can be used to visually capture and analyze data generated.
- **Step 7: Waste Elimination** – Determine which non-value activities can be reduced or eliminated from the process and redefine (update) the process upon waste elimination.

Kaizen Events

Kaizen events are targeted improvement activities with a specific goal in mind. The event is typically coordinated by an experienced facilitator who will lead a team in event planning, training, data collection, data analysis, and implementing actions. The objective is to eliminate waste, enhance operational performance, and increase customer value. Kaizen events support the concept of continuous, never-ending incremental improvement. It is understood that small and large process improvements, implemented over time, will positively impact operational efficiency and move the organization closer to operational excellence. Improvement must come from everyone in the organization, from line workers to top management. Sometimes improvements are small and instant while other times they are significant and intentional. Kaizen events focus on specific process opportunities that will likely be eliminated by a structured and disciplined approach to improvement.

Daily shop floor management can be looked upon as two separate activities, maintaining process control and driving productivity improvements. We can use Kaizen events to identify and eliminate problems that challenge our ability to achieve and maintain process control as well as to recognize process waste for elimination to improve operational efficiency. It is important to remember that successful events, reflecting sustainable improvements, are more likely to be achieved when management commitment and support are evident within the company.

Kaizen events are tailored to targeted improvements. For example, if cycle time reduction is necessary on a specific production line, an agenda is prepared to understand the process, take measurements to baseline the current cycle time, study the process for opportunities to reduce operator, machine, and wait times, discuss options to reduce observed waste and deploy action to realize improvements. An outline for conducting a Kaizen event can be found in Figure 4.2.5.

Kaizen Event

Figure 4.2.5   Kaizen event process.

Jishuken (Self-Study) Workshops

Productivity improvement is an essential survival skill for most manufacturing operations. It requires a keen awareness of one's operational performance and an understanding of where to deploy limited resources to maximize efficiency. Every organization must continuously engage in sustainable improvement activities relative to its level of operational maturity. In the early stages of a product lifecycle, opportunities exist to realize manufacturing capability. As processes mature, the operational focus turns toward maximizing process capacity. Once process capability and capacity are achieved, further optimization can be pursued by eliminating the obstacles to material and information flow. Jishuken workshops help to serve these continuous improvement objectives through self-study workshops.

Proactive organizations actively look for opportunities to improve operational efficiency. Unfortunately, opportunities for improvement are not always obvious to the untrained eye. Toyota has taken a head-on approach to process improvement by thoroughly understanding current operating conditions and working to actively identify improvements through Jishuken workshops which take a self-learning or "proactive" approach to process improvement. They are deployed to look for or highlight process waste, prioritize its impact, and systematically work to remove it in pursuit of greater operational efficiency. These self-study workshops take a learn-by-doing approach for continuous improvement. The workshops can also serve to develop employee awareness of waste and empower them to eliminate it, reinforcing the mindset of daily continuous improvement. We will briefly review three types of workshops focused on improving process capability, capacity, and optimization. An outline for conducting a Jishuken self-study workshop can be found in Appendix 6.

- ■ **Process Capability Workshop** – a capable process is one that consistently produces outputs that meet stated requirements and stakeholder expectations. This type of Jishuken workshop focuses on improvements needed to achieve process capability. The first step is to collect and analyze data to obtain a deep understanding of the current process and its shortfalls in not meeting output requirements. The root causes of underperformance are then reviewed in light of operating standards, system integrity, and compliance. Once the disconnects between the desired state and system failures are understood, actions can be taken to eliminate the disconnects and reinforce the operating system through updated standards and employee training. Once system integrity in meeting performance expectations has been confirmed, on-going system verification must occur to ensure system structure and deployment continue to satisfy stakeholder requirements.
- ■ **Process Capacity Workshop** – a capacity improvement workshop considers all constraints of time and motion required to maximize production output. The workshop deconstructs a process into its sequential work activities and further decomposes each activity into its fundamental elements characterized by a single action and time component. In essence, a process is broken down

into its most basic elements for analysis to provide insight into the waste and obstacles hindering material and information flow. In addition to discovering discrete opportunities for capacity improvement, this workshop takes steps to prioritize and implement actions to reduce process cycle time to within workshop targets.

■ **Process Optimization Workshop** – this workshop concentrates on creating a smooth, consistent flow of material and information throughout a product's value stream. It focuses on the disconnects and other obstacles to flow between value stream workstations and addresses the disruptions in material flow with techniques such as Kanban-based supermarkets and material buffers. The team creates a value stream map to reveal opportunities for improvement and defines actions to move from the current to a more desired state of operation through planning, plan execution, process monitoring, and control. Optimization is an on-going process in response to a continuously changing shop floor environment. It requires studying workflow for opportunities to deploy lean practices in order to incrementally enhance process robustness and performance.

Jishuken workshops can be a valuable addition to the shop floor improvement toolbox. They are most effective when an experienced facilitator works with knowledgable individuals focused on improving processes in their area of responsibility. These workshops are best when executed with structure and discipline. In addition to the actual time spent in the workshop identifying and implementing changes, the need to follow up on longer-term improvements will likely be necessary for the successful realization of workshop objectives. These workshops also serve as a valuable tool for developing individuals to exercise a lean mindset.

SIDEBAR: KAIZEN VS. JISHUKEN WORKSHOPS

Many know Kaizen as continuous improvement. Kaizen is often associated with a reactive approach to solving a specific problem impacting operational performance. A culture of continuous improvement involves the pursuit of perfection, knowing that it can never truly be achieved. As a mindset, this requires a proactive approach to working which moves beyond reacting to problems. This is where Jishuken workshops can add value.

When a process has been running well, but the team is looking for opportunities to implement further improvement, without clear direction or certainty of next steps, a Jishuken workshop can be initiated. A Jishuken workshop is a self-study activity intended to understand (or study) the current state of an existing process to identify and prioritize opportunities for improvement that enhance operational performance. It promotes the concept and mindset of incremental improvement by reinforcing the desired behavior by identifying and turning opportunities into action. See Figure SB-4.1

Topics	Kaizen Event	Jishuken Workshop
Goals & Objectives	> Intentional, specific & focused > Targets process weaknesses for elimination > Focused on solving a specific problem or implementing a chosen solution	> Broad, strategic & opportunistic > Identify process opportunities for improvement > Implement priority improvement opportunities identified through observation, data collection & process analysis
Vision - Organizational wisdom and knowledge transfer	> Employee ownership & skills development; > Incremental, significant process improvements	> Enhance management awareness, understanding & engagement in sustainable process improvement
Approach	> Targeted improvement in a specific area or process of interest. > Data collect to baseline the process & measure improvements	> Study the current state of the area or process for improvement.; analyze process losses > Deep-understanding of process fundamentals > Establish baseline performance & highlight opportunities for improvement
Frequency (Typical)	Continuous / Weekly / Monthly	Quarterly / Semi-annual / As Needed
Duration	Several hours or days to complete the event	> Several days for the workshop > Several weeks to complete improvements
Improvement Leader	Team, department or area Leader, where improvement is occurring	Experienced facilitator / coach internal or external to the facility
Team Composition / Participants	A small team of people working together to realize area improvement targets	Several teams of 2 to 3 people (managers & employees) working together to identify opportunities & implementing solutions for process control & improvement.
Management Role	Management endorsement & support	Direct (high-level) management involvement

General Tips:
Confirm conformance to standards
Look for & highlight process abnormalities / deviations to standards
Define & monitor key process indicators
Determine cost & productivity impact
Deploy improvements in all applicable areas
Strive for little to no additional investment.

Figure SB-4.1   Kaizen events vs. Jishuken workshops.

Productivity Improvement Tools and Techniques

There are a plethora of different tools and techniques available to help improve manufacturing productivity. Some of these tools and techniques are based on technological advances, some exploit the use of software, while others leverage a more traditional approach to data gathering and analysis. Regardless of the method used, a good tool or technique will help individuals and teams achieve their targeted objectives easier and faster than not using them at all. Although there is no clear line between a method, tool, or technique, the following pages are intended to provide an overview of some of the more common and practical methods, tools, and techniques used for improving daily shop floor management.

Go, See, Observe, Understand, and Confirm

One of the best, most effective, and readily available tools at your disposal is your eyes for personal observation. Observation is an extremely effective way to collect factual data for analysis. When engaging in improvement activities, it is important to periodically stop and observe for 10, 20, 30, or more minutes to learn first-hand what is actually happening at a workstation or production area. This is a powerful technique in that it creates awareness, understanding and generates facts for the observer. In an area or operation of interest, take time to observe the work being performed. During your observations, consider:

- What activities are people doing? Is there anything "unusual" in what is happening?
- Are people's actions consistent with area standards, procedures, and work instructions? Look for working patterns.
- How are people handling materials and interacting with equipment?
- What disruptions to workflow do you observe and how are people managing those disruptions?

Take time to stop and observe shop floor activities frequently. Note what you see and with who you talk. These references can become important as issues arise and problems are being solved. More on this topic of observation can be found in the sidebar: tips for effective observation.

SIDEBAR: TIPS FOR EFFECTIVE OBSERVATION

- Go to the gemba frequently to observe and talk to people about what they are doing and what's happening.
- Tell the Line Leader/Supervisor why you are in the production area.
- Find the best spot to observe without disrupting production operators or workflow.
- Observe the line running; does the assembly process continuously flow? If not, why not? *What things are causing disruptions in operator movements*

or material flow? Are operators waiting on machines or are machines waiting on operators?

- Stop and observe one activity (for 10+ minutes), and look for work patterns, disruptions, or a lack of a pattern where one should exist.
- Note any deviations from operating standards. Look for a change in people's routines. Are they performing to standard? If not, why not?
- Does the actual cycle time observed match the planned cycle time? *If not, why not?*
- Identify and observe the bottleneck; can its cycle time be improved if needed?
- Write down what you observe as you observe it. Try to be as specific and quantitative as possible (e.g. how much time an activity takes or the number of units rejected).
- Observe how operators interact with their equipment; is it seamless, graceful, or do they struggle?
- Does the operator leave the work cell or area? If so, why?
- Observe an operator handling material. Can a small change in material placement or positioning make it easier? Remember, repetitive routines can be exercised thousands of times a shift. Minor adjustments make a difference!
- Pay attention to the small stuff, the details; every second counts.

Workplace Design/Ergonomics

Managers must manage while workers must work. This is best accomplished when the environment within which these things are happening is suitable and conducive to doing so efficiently and effectively. This means that the workplace should be considered as part of the continuous improvement paradigm. This can serve manufacturing by treating workplace layout and design as an opportunity to further lower the time and effort it takes people to complete their work tasks and activities.

Look to eliminate waste in the workplace, such as excess operator motion, and the need for material transport, through better utilization of floor and storage space. Minimize maintenance efforts through the availability of proper tooling. Reduce the discomfort of working in one spot all day with chairs and fatigue mats, periodic exercises, and frequent breaks. Create easy-to-use interfaces between man and machine through motion studies and follow-up action. In doing so, it is important to respect employee safety and enhance workplace ergonomics. Employee comfort helps employee retention and reduces mistakes.

Process Mapping

Process mapping is used in manufacturing to visualize process workflow. It is sometimes called flowcharting or workflow diagramming. The preparation of a process map helps to create awareness and understanding of work activities and their corresponding sequence of events. Mapping is typically the first step in process improvement since a well-documented map can highlight potential process concerns, bottlenecks, and inefficiencies leading to opportunities for productivity improvement.

There are different types of process maps including a high-level SIPOC which identifies Suppliers, Inputs, Process activities, Outputs, and Customers. Functional Deployment mapping details process activities and their corresponding sequence. Swim lanes can be added to a functional deployment map to visualize the department or positions responsible for completing the activities defined. A Spaghetti diagram is also a form of process mapping that illustrates the movement of people and materials in an area or process of interest such as a warehouse or work area.

There are many books and articles that cover the topic of process mapping that will not be repeated here. However, it is important to remember that it is a valuable technique to employ when looking to augment an individual or team's understanding of a process targeted for improvement.

Six Thinking Hats

Six thinking hats is a creative problem-solving technique that uses a team role-playing approach that considers six different perspectives (symbolized by colored hats) on a topic of interest. It is designed to enhance team creativity, brainstorming, and evaluate ideas with minimal conflict. When facilitated properly, it is relatively simple, practical, and fun. The colored hats represent the following perspectives during a team session:

- **White Hat** – focuses on facts, figures, and objective information.
- **Red Hat** – provides an emotional/feeling-based input.
- **Black Hat** – injects logic into the thinking/brainstorming process.
- **Yellow Hat** – offers a sunny, positive, and constructive viewpoint.
- **Green Hat** – shares new, novel, and creative ideas with the team.
- **Blue Hat** – facilitates, oversees, and controls the thinking process. Determines who wears what hat.

Role playing by the team "Thinkers", as represented by the hats, can take several forms:

- Thinkers can wear or remove a hat.
- Facilitator can ask a Thinker to wear or remove a hat.
- Thinkers all put on the same hat for a given period.
- Thinkers are each assigned a different hat to wear for a discussion period.
- Thinkers wear hats they do not typically wear.

More information on this topic can be found in a book titled *Six Thinking Hats* by Edward De Bono [2].

Workload Balancing

Workload balancing in manufacturing is the leveling of work in process. This is where all process workstations are targeted for the same amount of work, creating a

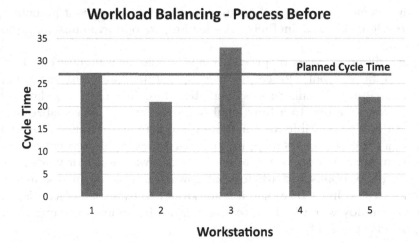

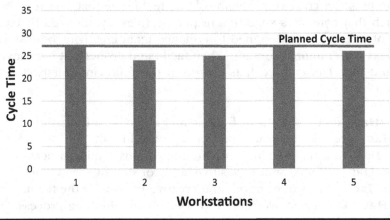

Figure 4.2.6   Workload balancing – before and after.

consistent and even workflow. This can be accomplished by studying the cycle times of each workstation within a process and comparing the results, relative to each other and the planned cycle time, looking for significant differences. If inconsistencies are evident, first look to eliminate process waste, followed by reassigning tasks between workstations to create a more even work distribution among operators and machines. The objective is to achieve a similar cycle time between workstations consistent with the planned cycle time for a production line. See Figure 4.2.6. Workload balancing helps facilitate process stability, operational efficiency, and the steady flow of products to the customer.

Production Leveling

Production leveling is a technique for reducing production unevenness to manufactured goods at a more uniform and predictable rate of output. When customer demand

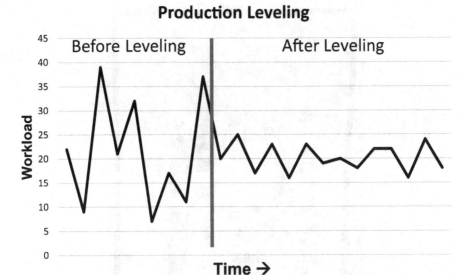

Figure 4.2.7   Production leveling.

is consistent, production leveling is easy. Unfortunately, this is often not the case. The flow of work through production will likely be unpredictable, disruptive, and inefficient if the production scheduling team simply released orders to the shop floor upon receipt. To even the flow of work through the factory, the logistics team will specify a lead time for filling customer orders, in order to prepare a build schedule several days or weeks in advance that will balance the workload, creating a more stable, efficient and predictable rate of production output.

Production leveling can refer to the leveling of product volume, product type or product mix, all of which are closely related. A production schedule can be prepared to help visualize and release work orders at a controlled pace. Production inventory levels can also be used to regulate difficult-to-control variations in customer demand. The concept behind production leveling is to average out longer-term production demand and maintain an inventory proportional to the variability of demand to achieve and retain operational stability. See Figure 4.2.7.

Yamazumi Charts/Boards

Yamazumi is a Japanese word that means to *stack up*. Toyota uses Yamazumi work balance charts to visually present the content of a series of work tasks performed on a production line. The charts are composed of stacked bars representing key cycle time contributors and graphically display opportunities for process optimization. The objective of these charts is to facilitate operator work balancing by isolating, visualizing, and targeting non-value-added work for elimination.

Process tasks of each operator working in a team are individually represented by a stacked bar. One bar reflects the repetitive work tasks (cycle) of one operator.

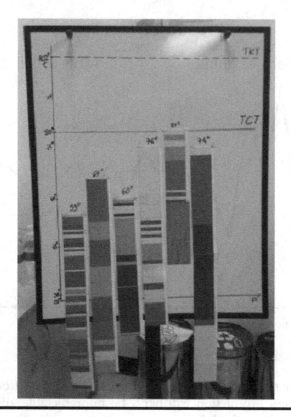

Figure 4.2.8 Yamazumi chart.

The stacking of sequel tasks in each bar is classified (using color) as value-added and non-value-added (waste). Red is typically associated with non-value-added tasks. The mean duration of each task within a stacked bar is represented by the bar size. The thicker the bar, the more time it takes to complete the task. The stacked bars of all operators working together in production are then placed next to each other on a board or chart relative to the planned cycle time. The Yamazumi chart illustrates which operators are underutilized and overloaded relative to planned cycle time and can provide a mechanism to quickly identify and rebalance a process when the planned cycle time changes. It is a great tool for visualizing delays, waste, and constraints. An example of a Yamazumi chart can be found in Figure 4.2.8.

Buffer Management

A manufacturing buffer is a defined amount of material such as raw material or work in process (WIP), staged at a workstation, to enable a line or value stream to maintain a smooth, steady workflow through the process. It is commonly used in manufacturing to compensate for the natural disruptions (variation) in material flow when transforming raw material into finished goods. A buffer is a tool to help maintain production flow interrupted by process abnormalities, adjustments, deviations and periodic work. Buffer levels can be changed to regulate the movement of material through a system. Buffers can "buy" time to solve problems and eliminate obstacles to process flow without impacting customer deliveries. Increasing buffers should be done with the clear intent of lowering them back to a "normal" or acceptable level for continuous flow without compromise.

One way to establish a buffer in production is to observe the process in operation and look for times when operators are waiting for units from an upstream worker or machine to perform their tasks. To maintain flow, continually add units to the buffer of waiting operators until their wait time is eliminated and steady flow is achieved. The buffer quantity needed to establish flow must be determined for each product variant based on the cycle time of the task performed at each workstation.

Buffers serve as a reference or baseline for process improvement with the objective of reducing buffer limits to the lowest possible level necessary to hedge against unacceptable process variation. Steps should be taken to understand obstacles to flow within a workstation, production line, or value stream and strive to optimize the flow of material and information by understanding the cause of each obstacle and implementing actions to eliminate their impact on output performance.

Best Practice Sharing

Sharing best practices within an organization is a smart way to facilitate continuous improvement and promote a learning culture. Operations mature faster when employees seek, evaluate, and adopt the good/best practices of others, to their advantage. Deploying a best practice sharing program within a company can bring everyone up to the highest standard of performance quicker than getting there on their own. The challenge is to identify existing practices that can benefit others and eliminate the obstacles to sharing and implementation. Sharing best practices can be as simple as having subject matter experts (SMEs) and experienced employees explain what they did, why they did it, and how they did it. Learning from the knowledge and experience of others can accelerate operational maturity and provide a competitive advantage when done well. The following are some best practice sharing tips:

- Use the expertise of your Subject Matter Experts and top performers to identify and share best practices.
- Eliminate the silos that create obstacles to sharing good/best practices. It's not an individual or plant-to-plant competition. It is about exploiting the collective wisdom of everyone for the benefit of the entire organization.
- Incentivize good/best practice sharing through encouragement, recognition, and rewards.
- Schedule regular good/best practice sharing meetings between groups; dedicate time for sharing best practices.
- Create a good/best practice sharing site for storing useful information and relevant documentation.
- Share skilled people's time with other departments or locations to exploit their expertise for maximum company benefit.
- Make it simple, easy, and rewarding for people to share their knowledge and best practices.

Lessons Learned

There are several definitions for the concept of lessons learned. One used by the national aeronautics and space administration (NASA) is "a lesson learned is knowledge or understanding gained by experience. A lesson must be significant and applicable in that it identifies a specific design, process or decision that reduces or eliminates the potential for failure". Sharing lessons learned among individuals and teams helps prevent organizations from repeating the same mistakes and allows them to benefit from best practices. Application of lessons learned can improve operational efficiency and competitiveness.

Lessons are learned all the time. It is important to periodically stop and recognize when these opportunities occur and capitalize on them through documentation and sharing them with others. Whenever starting a project or similar activity of the past, consider reviewing lessons learned at the beginning so that the mistakes of the past can become a learning opportunity for those moving forward. A structured approach to identifying, capturing, storing, and sharing lessons learned within the organization will increase their impact and sustainability.

Key Points

- Efficiency improvements come from empowering people at all levels of the organization.
- Engage in continuous productivity improvement using innovative thinking and an evolutionary approach.
- Observation is used to identify improvement opportunities while experiments are conducted to evaluate and validate process changes to elevate process performance standards.
- Successful failure leads to learning.
- Productivity improvement is "easy"; all you need to do is lower your production costs per unit or increase your company sales without increasing fixed costs.
- Process control and continuous improvement should not be treated as separate, specialized, or unique activities, they must be integrated into the daily work routines of every employee and measured periodically to ensure they are happening.
- Proactive companies remain vigilant and constantly look for opportunities to implement continuous improvements to hedge against the threats posed by dynamic and changing market conditions.
- Embrace employee inputs and ideas. Take a structured approach in doing so.
- Create a culture in which sharing ideas is encouraged. Incentivize people to share.
- Waste reduction and process flow improvements are the driving forces for elevating manufacturing performance.
- Kaizen events involve conducting small experiments, analyzing results, making changes, monitoring activities, and initiating adjustments to achieve sustainable improvements.
- Jishuken workshops are used to analyze the process in detail and identify opportunities for significant improvement.

Key Points – Continuous Improvement

■ Continuous improvement is the act of improving process efficiency by eliminating waste and removing obstacles preventing the uninterrupted flow of materials and information through the value stream.

■ Continuous improvement proactively looks for potential problems and waste for elimination and requires countermeasures that prevent these issues from reoccurring.

■ Continuous improvement must occur throughout an organization, focusing on product, process, systems, and employee development.

■ Incremental improvements are made daily when issues and opportunities are discovered during the execution of daily work routines.

■ Incremental improvement should be driven by deliberate and gradual change rather than rapid "fixes" that may help in the short term, but become longer-term chronic problems.

■ Continuous improvement is everyone's responsibility and should be an integral part of employees' daily work routines.

■ An attitude of continuous improvement must become a mindset and habit of every employee, making it an integral part of the company culture and DNA.

Sources

[1] Rother, M. 2014. *Improvement kata handbook* V23.0. http://www-personal.umich.edu/~mrother/Handbook/Full_IK_Handbook_v32.0.pdf

[2] de Bono, E. (1999). *Six thinking hats*. Back Bay Books, Little, Brown and Company, MICA Management Resources, Inc.

4.3

Agile Methods

If you want to be fast and agile, keep things simple. Speed isn't the result of simplicity, but simplicity enables speed.

~Jim Highsmith

Objective: Strive to remain agile in how we think and act every day.

Overview

Agile originally started as a project management methodology in response to a need to be more flexible and responsive to changing customer requirements during the development process. It requires constant stakeholder engagement and collaboration throughout the project development phases. Agile was born from a Manifesto compiled by software industry leaders in 2001 to articulate better ways to develop software. The agile manifesto is composed of twelve principles as follows:

- Our highest priority is to satisfy the customer through early and continuous delivery of valuable software.
- Welcome changing requirements, even late in development. Agile processes harness change for the customer's competitive advantage.
- Deliver working software frequently, from a couple of weeks to a couple of months, with a preference to the shorter timescale.
- Business people and developers must work together daily throughout the project.
- Build projects around motivated people. Give them the environment and support they need and trust them to get the job done.
- The most efficient and effective method of conveying to and within a development team is face-to-face conversation.

DOI: 10.4324/b23307-25

- Working software is the primary measure of progress.
- Agile processes promote sustainable development. The sponsors, developers, and users should be able to maintain a constant pace indefinitely.
- Continuous attention to technical excellence and good design enhances agility.
- Simplicity – the art of maximizing the amount of work not done – is essential.
- The best architectures, requirements, and designs emerge from self-organizing teams.
- At regular intervals, the team reflects on how to become more effective, then tunes and adjusts its behavior accordingly.

These principles are no longer unique to software since they were quickly adopted by other disciplines and industries looking to enhance their development capabilities. An agile approach embodies a mindset where specific practices are selected and applied based on need.

Application of Agile Practices

The application of agile practices on the shop floor is not typically a common occurrence since volume production is built on standard processes and procedures to achieve consistent and predictable output. Normally, teams can plan and manage on-going operational activities with clearly defined and documented requirements. As the uncertainty of these requirements increases, process stability starts to waver, leading to operational inefficiencies which can become costly and time consuming. Occasionally, a request or event occurs outside the operational norm, causing a process disruption that may benefit from an agile approach for effective resolution.

Agile practices may be most beneficial to manufacturing in a product and process development capacity due to the new, unique and difficult requirements that are likely to be occasionally encountered. Highly uncertain work is characterized by new designs, high complexity, frequent changes, and lots of experimentation and problem-solving conducted within a never-done-before environment. It involves significant risk. In this scenario, agile is a good approach since it's a method that you can use when you don't know exactly what to do. It requires a continuous feedback loop, with key stakeholders, as the product and corresponding requirements continue to evolve during the development cycle. It's one of the best ways to manage highly uncertain work while navigating risk.

When a high degree of uncertainty exists, risk can be mitigated via small increments of work defined in work packages. Small increments of work require frequent verification and adjustments to obtain a rapid and accurate understanding of real-time stakeholder requirements and final deliverables. Groups that adopt an iterative and incremental life cycle approach respond to changes more easily, reducing waste and rework. This is due to the use of short feedback loops, frequent process adaptions, continuous work re-prioritization, progress plan updates, and regular demonstrations of effectiveness.

Predictive, Iterative, and Incremental Life Cycles

The predictive life cycle is the traditional approach to development using an upfront planning and "single pass" requirements realization process. This lifecycle model is typical of a continuous manufacturing operation (vs. Customized) where the work is planned in detail based on known requirements and constraints. A typical predictive life cycle flow can be found in Figure 4.3.1.

An iterative life cycle approach involves frequent feedback to continuously refine work-in-progress prior to final delivery. Regular stakeholder involvement occurs with changes incorporated at periodic, logical intervals. Iterative characteristics include high complexity, frequent changes, and differing stakeholder views based on feedback from the latest activity, information, and results. This iterative life cycle of activity may be seen in production start-up activities and deployment of strategic projects within a manufacturing environment. Figure 4.3.2 reflects an Iterative life cycle.

An incremental approach focuses on frequent deliverables for immediate stakeholder evaluation and potential application. This is ideal when validating a significant change to manufacturing operations. Presenting stakeholders with a new or revised plant layout to accommodate a new product line, before starting work, can employ an incremental approach. The feedback may result in adjustments or changes before additional time, work, or money are invested. This can reduce potential rework and avoid stakeholder dissatisfaction. As another example, an incremental approach may be applied to product development, where successive prototype builds by the manufacturing team allow for early and frequent product manufacturability and assembly feedback for process development and refinement.

A combination of both **iterative** to *refine work items* and **incremental** to *deliver frequently* can be deployed as needed. Activity deliverables are elaborated frequently during an iterative approach. Key stakeholders are involved in on-going activities while changes are integrated real-time. Risks are managed and cost controlled as requirements and constraints evolve.

Agile approaches are often used because requirements are expected to change. Feedback influences the next actions taken in the development process. It's not

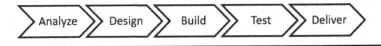

Figure 4.3.1   Predictive life cycle.

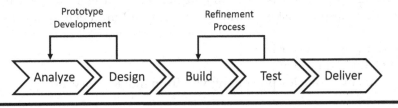

Figure 4.3.2   Iterative life cycle.

unusual for incremental deliveries to reveal unexpected or misunderstood requirements. Incremental deliveries that adapt and align with stakeholder needs can occur in different ways. Progress is measured by incremental deliverables that provide functional value to stakeholders. In iteration-based agile, iterations are conducted in timeboxes, *of equal duration*, to achieve timely deliverables.

A hybrid life cycle combines elements from the different life cycle models such as predictive, iterative, and incremental approaches. The hybrid approach can be applied when significant uncertainty, complexity, and risk exists in development, followed by a phase with clearly defined and repeatable steps. An example of this might be a pharmaceutical company requiring FDA approval (predictive) following process development (agile). In addition, elements such as short iterations, daily stand-up meetings, and retrospectives can be employed while an organization is incrementally transitioning from a predictive to agile approach.

For most organizations, the transition to agile will take time. Agile techniques look and feel different from those of a predictive model. For organizations, especially large ones, a gradual transition is recommended. Start the transition by adding more iterative techniques to enhance learning and team alignment. Consider the following.

- Follow up with more incremental techniques to accelerate value and stakeholder "return on investment" (ROI).
- Start the transition with lower risk projects.
- Increment with more agile techniques, applied to more complex activities, once teams demonstrate confidence with a hybrid approach.
- Sense the team's willingness to adapt and embrace more transitional changes before moving forward.

Planning is the common theme among all these life cycle approaches, the difference being how much and when planning occurs. Predictive involves upfront requirements definition while planning drives work. Iterative relies on prototype and proof outputs to drive modifications to initial plans. Incremental applies to delivering successive subsets of overall outcomes. Agile requires team planning and re-planning as more information becomes available from reviews and frequent deliveries. In essence, task deliverables or project characteristics will determine the best life cycle approach for application.

Daily Stand-Up Meetings

The daily stand-up meeting is a "15-minute" event for a team to plan the next 24 hours of work. Questions asked during a typical stand-up meeting may include:

- What did you do yesterday?
- What will you do today?
- Are there any obstacles preventing you from moving forward?

These meetings help teams understand the status of specific tasks and what needs to be completed next. If roadblocks exist, the team can work together to understand

and eliminate them. When conducting a meeting, it's important to keep it focused on the agenda and within the allocated time. Don't start rambling or problem-solving, you don't have enough time. Allow everyone invited to contribute to the discussion and pick the best time for everyone needed, to attend. Strive to hold the meeting at the same time and place every day. Promote an open environment to share issues and obstacles with team members. Consider the following tips for holding short "stand-up" meetings.

Tips for conducting a stand-up meeting:

- If a meeting update becomes lengthy, schedule it for a later discussion.
- Respect everyone's time; keep updates clear, quick, and concise.
- If feasible, everyone *should* stand up during the daily meeting.
- Only one person speaks at a time. Pass around a "speaking token" to facilitate this activity.
- Schedule a separate, more structured problem-solving approach, when a problem reoccurs.
- Start and end stand-up meetings on time. Create deterrents for those who arrive late (i.e. penalty jar).
- Use a stand-up meeting agenda and exercise a structured/formatted approach.

Retrospectives

A retrospective is a process for capturing lessons learned relative to what went right and what went wrong. It's a meeting held at the end of an activity or deliverable to reflect on what happened and highlight actions for making improvements moving forward. Retrospectives encourage participants to deliberate about enhancements for the next tasks or deliverables. Team members are encouraged to provide feedback, share their views, and agree on needed adjustments. Retrospectives provide the opportunity for teams to look back and see how they can improve while being a catalyst for organizational change. It's important that a retrospective focus on process, not people.

Co-Location

Co-locating certain groups of employees helps bridge the communication gap. People working within the same office area tend to communicate more, improving productivity. When communication becomes a critical part of daily work, efforts should be made to bring people together, even if only for a brief period, especially when agile decision-making is required to advance a project or complete a critical assignment. If this is not possible, schedule regular communication sessions so information can be exchanged, questions answered, and decisions made. Only spend the needed time to communicate since anything more is waste. A 30-minute meeting that takes less time should end when meeting objectives are met. A 30-minute meeting that requires more time may not have been properly planned. Strive to create and facilitate a comfortable environment to communicate.

SIDEBAR: AGILITY RESPONSE

When work priorities exceed the time and resources needed to meet schedule demands, it's time to deploy agility measures. One agility technique proven to optimize time and resource availability is the daily stand-up meeting. At the start of each work day or shift, all available resources, within a department or function, meet as a team for 10 to 15 minutes to review the highest priority activities, consider resource availability and expertise, and assign work tasks accordingly. The following are some rules of engagement to consider:

- A qualified task master is assigned to lead the agility effort.
- Daily (or more periodic) stand-up meetings are held (same time and place).
- At each meeting, people must be prepared to report:
  - ✔ What was accomplished since the last meeting.
  - ✔ Highlight any issues or obstacles preventing progress.
  - ✔ Next steps.
- Prioritized tasks must be visualized (on a board) for everyone to see during the discussion.
- Status of prioritized actions is updated in real-time, during the meeting.
- Everyone should have an opportunity to speak.
- Daily tasks are discussed, and assigned, once the current state has been established.
- All prioritized tasks, assigned resources, must be completed before those resources become available for reassignment, unless the task was put on temporary hold for an external issue beyond the facility's control.

CASE STUDY: MANUFACTURING AGILITY

Although not inherently obvious to many, agility is a cornerstone of Lean manufacturing. I observed a good example of agile in application during the 2020/2021 pandemic. The pandemic challenged many manufacturing facilities to continue "normal" operations in light of significant employee absenteeism due to the corona virus illness and the need for other employees to isolate when exposed to the virus. This situation created shortfalls in the workforce as manufacturing teams struggled to maintain production demands. In response, the management team took a scrum approach to production scheduling by meeting every day (e.g. stand-up meetings), sometimes every shift (at the height of the pandemic) to assess workforce availability and make adjustments to the production schedule based on delivery priorities and available employees. In the absence of a full workforce, certain production lines were shut down and employees were reassigned to other lines in order to satisfy the highest priority deliverables.

The need to practice agile manufacturing continued even after more employees started returning to work. Although vaccinations reduced the absenteeism rate in the United States, slow roll-out of the vaccines in other countries caused further delays in obtaining material from suppliers abroad. During this same time, a highly contagious variant of the virus took hold, generating a subsequent wave of employee isolations and illnesses. This resulted in certain global suppliers slowing or shutting down their production of critical components needed for making automotive electronics in the United States and elsewhere. Once again, automotive electronic plants responded, in this case, to a parts shortage, by shutting down affected product lines and re-tasking people when there was less work to do. These are just a few examples of the many different scenarios manufacturing teams faced in the wake of the pandemic.

In the highly competitive manufacturing industry, maintaining the lowest possible cost per unit produced is key to survival. This is difficult enough under stable manufacturing conditions. Unfortunately, when these conditions start to change and change frequently, normal operating practices must be supplemented with agile methods to address the unique and often unpredictable circumstances that occur in an increasingly complex work environment.

Key Points

- Agile is as much a way of thinking as it is a way of working.
- The agile methodology is what you do when you don't know what to do.
- Hybrid life cycles can be crafted based on project and work risks.
- Select an agile project life cycle to employ based on current business needs.
- Agile is presented as a *framework* which should be tailored by teams to deliver stakeholder value.
- Agile projects do not use standardized or repeatable processes. What worked in the past may not work for the present or future.
- Agile is about adapting existing practices to suit the needs of the project and organizational work style.
- The objective is not to be agile for the sake of being agile, it's to increase customer satisfaction and improve business outcomes when expectations are new and uncertain.

Source

The Guide to the Project Management Body of Knowledge, sixth Edition. Published by the Project Management Institute, 2017 (WWW.PMI.org)

4.4

Shop Floor Optimization

Optimization is reflected in the capability of a process or system to consistently meet stated requirements at the lowest possible cost and time.

Objective: Make the most cost-efficient and effective use of manufacturing resources in meeting stated requirements.

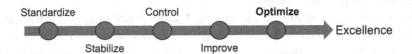

Overview

Manufacturing optimization is realized when the continuous flow of materials and information produces consistent and predictable output performance at the lowest possible cost and time. The challenge is to maintain this state of existence indefinitely and under the constant threat of manufacturing disruptions due to people, equipment, materials, methods, and environmental instability. Manufacturing optimization is difficult to achieve and even more difficult to maintain as factors influencing it are in constant flux and must be continuously monitored and managed.

Manufacturing system optimization is a catalyst for manufacturing excellence. The purpose of optimization is to achieve the maximum benefit, functionality, and effectiveness from a person, place, or thing, relative to a set of criteria or constraints. In manufacturing, optimization often focuses on the efficient application and harmonization of people, equipment, materials, methods, and the environment. Holistically, optimization of manufacturing operations requires knowledgeable and skilled people, standardized processes, reliable equipment, consistent materials, and a controlled

DOI: 10.4324/b23307-26

environment, among other things. In an optimized system, these elements work in harmony to create an efficient and uninterrupted rhythm of work, producing desired results.

Characteristics reflective of an optimized system don't come easily. They require planning, execution, monitoring, and continuous control to achieve and maintain a stable and efficient state of operation. The pace of production flow, within an optimized state of control, is likely to vary within limits, due to changing customer demands and supply chain dynamics. In response, periodic adjustments to process inputs and activities will likely be required. Optimization is a relatively simple concept to understand but a lot more difficult to realize under increasingly complex and turbulent working conditions. It requires highly trained people able to adjust their activities in response to changing circumstances. Let's consider some of the factors impacting manufacturing optimization in this chapter.

Knowledgeable and Skilled People

An optimized process requires knowledgeable people who know what to do and when to do it. Cross-training people, in performing multiple job responsibilities, increases workforce flexibility and agility, in response to changing operating conditions. Enhancing people's knowledge and skillsets creates a competitive advantage, facilitates professional growth, and opens the door for employee advancement within the company. This approach also helps to maintain a stable workforce and retain valuable know-how and talent.

Developing people's ability to identify and solve problems, at the lowest possible level of the organization, frees up employees with specialized skills to work on more complex issues. Engaging people in idea generation, qualification, and implementation facilitates continuous improvement, which is another essential component of optimization and operational excellence. Training everyone in problem-solving and continuous improvement, while empowering people to use these skillsets every day, creates tremendous potential for growth and advancement of company objectives.

The availability of work instructions and standard work routines provides people with clear direction and expectations to efficiently complete their work activities while allocating time to respond to the unexpected events that frequently occur when performing daily work routines. Motivated individuals, such as production line operators, can be trained to troubleshoot and maintain their operating equipment within manufacturer specifications. Supervisors can be developed to manage, coach, and mentor employees while all employees can be trained to solve problems that occur within their scope of responsibility, allowing them to make the best and most efficient use of their time, at all times. Topics to consider when managing people:

- Hire the best people available. People are the reason companies are successful.
- Unreliable and unmotivated people don't make good employees.
- Cross-train people to optimize company flexibility and agility.
- Pick the best managers to manage people. Good people don't necessarily make good managers.

- Diversity brings different perspectives to problem-solving and decision-making.
- Create an environment in which people want to work.
- Plan and develop the best people.

Cooperation and Collaboration

Collaboration is necessary because optimization needs the collective knowledge and skills of an experienced workforce working together to experiment, learn, and achieve sustainable improvements. Effective collaboration can only occur when cooperation exists. Collaboration allows individuals and cross-functional groups to work cooperatively together for the benefit of the whole. Optimizing a manufacturing process that crosses functional boundaries is best achieved when everyone is working together toward common goals and objectives. This is why, organizations that fail to deploy their strategy and policies throughout the company struggle to optimize their operations since employees are likely to pursue different paths based on *their* own priorities and *their* own decisions on the best way to do something. Employee priorities must be aligned throughout the organizational hierarchy to facilitate cooperation, efficiency, and collaboration; essential elements of optimization.

Available and Reliable Equipment

If your equipment is not available when you need it, your process is not running at its optimal level of performance. Equipment maintenance is easy to forget when all is running well and then, suddenly it's not! Total productive maintenance (TPM) is a price you pay for manufacturing optimization. TPM is about maintaining equipment to run at peak performance. This requires that you service equipment according to the original equipment manufacturer's (OEMs) directions and specifications. A TPM program can be started by following the equipment manufacturer's maintenance schedule and building upon the documented routines based on local usage and lessons learned through corrective and preventive maintenance activities. In certain circumstances, you can improve upon the robustness and reliability of equipment by observing their wear characteristics and replacing certain parts with more wear-resistant materials and robust components. These small actions all contribute to achieving an optimal state of equipment performance.

The actions described are no different than taking the family vehicle for service. If you don't want family members to experience an unexpected breakdown or unsafe driving conditions, you respect the vehicle's maintenance schedule and, on occasion, enhance vehicle reliability with better tires, synthetic oil, or a stainless-steel exhaust system. When considering the cost of a vehicle or manufacturing equipment, common sense dictates regular maintenance. However, it may not be just about keeping equipment running at peak performance, it may be more about ensuring equipment availability to meet customer delivery dates since the consequence of not doing so may have a significantly negative impact on the bottom line.

As discussed in previous chapters, quick recognition and response to failing equipment or repair of broken equipment as well as maintaining an inventory of critical spare parts will significantly reduce unplanned downtime and help maintain continuous process flow, a hallmark of manufacturing optimization and excellence. An even more progressive approach is to engage in predictive maintenance by collecting and analyzing equipment data, looking for abnormal behavior, or deteriorating performance trends, in order to anticipate potential failure modes and address them before they adversely affect production stability.

Tip for Managing Equipment

- Prepare a TPM schedule and stick to it!
- Select equipment based on value; consider short- and long-term costs, reliability, maintenance, ease of use, spare parts availability, etc.
- Monitor equipment downtime. Target improvements to equipment with the most (costly) downtime.
- Encourage and train operators to take ownership of their equipment. Prepare them to perform minor maintenance and repair.
- Budget for equipment replacement when they no longer meet manufacturing expectations or operating requirements.

Materials Management

Optimization requires that quality material is available on time and in the quantities needed to support the production schedule. This entails reliable suppliers, compliant material, accurate inventory control, good scheduling, and the operational discipline to ensure raw material and components reach their intended shop floor destination on time and without incident. If multiple suppliers are used for the same material, monitor the differently-sourced material for significant variations in form, fit, or function. Strive for First-in, First-out (FIFO) material control for better traceability and to avoid material obsolescence. Periodically verify that material is being controlled to manufacturer recommended practices.

Too much or too little material received at any one time can disrupt the supply change when striving to achieve or maintain an optimized process. Strong partnerships with key suppliers are essential for maintaining process integrity since the quality of any system is only as good as its weakest link. The following are some tips for material management.

Tips for Managing Material

- Minimize your supply base – a larger supply base than needed requires more resources, time, and money to manage.
- Favor suppliers that are located close to the manufacturing facility, especially if transportation costs are significant.
- Maintain good supplier relationships. Treat key suppliers as manufacturing partners.

- Clearly communicate your requirements through drawings and specifications.
- Closely monitor and work with suppliers not meeting expectations. Attitude is everything!
- Manage material inventories to a plan.
- Automate material reordering.
- Actively manage material aging to avoid excess and obsolescence.

Practices and Procedures

Practices and procedures are part of the organizational structure required for optimization. Discipline demonstrates the behavior needed to achieve a consistent and predictable shop floor operation and accountability of shop floor personnel ensures optimization can be realized. The practices and procedures of an organization define the operating standards and the discipline required to implement those standards as documented. Standards articulate how people are expected to work, equipment is expected to be maintained, materials are expected to be managed and the environment is expected to be controlled. In this context, practices and procedures create the framework for shop floor management and optimization.

According to Merriam-Webster.com dictionary, optimization is defined as "an act, process, or methodology of making something (such as a design, system, or decision) as fully perfect, functional, or effective as possible". It's with this same intent that organizations define practices and procedures for functional teams and employees to follow. A practice or procedure defines a way of doing something. They articulate a way of working and can serve as a guidepost or roadmap for operational stability, control, improvement, and optimization. In short, if you want to optimize your shop floor activities, it's best to provide a set of practices and procedures that will guide the workforce to an ideal state of operation. Given the right direction, proper training, sufficient resources, and effective management, optimization will become part of the journey to excellence.

Tips for Managing Practice and Procedure

- Tailor your practices to align with your strategic objectives.
- Clearly define and communicate the practices and procedures employees are expected to follow.
- Ensure awareness and understanding of practice and procedures for proper deployment.
- Continuously refine your practices and procedures, based on experience, to align with your journey of optimization and excellence.
- Periodically confirm compliance to key practices and procedures; take action, when needed, to address significant deviations.

Operator Movement Optimization

If continuous material and information flow, to a specified takt time, is key to manufacturing optimization, then the assembly operator must become an integral part of that rhythm. The rhythm is dictated by production line design and influenced by standard work procedures as well as planned cycle time. Optimizing operator movement to work within the cycle time of the line's intended rhythm requires attention to an operator's movements including their arms, hands, head, and feet. Understanding production line workflow must include the flow of an operator and how their body movements interact with their working environment including equipment, buffers, and line-side raw material storage. When capacity limits of a product line require it to continuously run at an optimal rate of output, operator movements may need to be choreographed to consistently meet line performance expectations.

Material Flow Optimization

When looking to optimize the flow of materials, there are three areas to consider, the flow of raw material to the production floor, work-in-process (WIP), and finished goods. The flow of these materials is influenced by many factors including storage location, distances to destination, material container type and sizes, packaging materials, machine cycle times, material transfer process, and transport equipment such as conveyors, carts, forklifts, and people. Unobstructed material flow, along the value stream, is a significant attribute of process optimization and, as such, is a prime focus of the value stream mapping/design methodology that will be discussed later in this chapter.

Environmental Optimization

When environmental conditions such as temperature, humidity, or air cleanliness are needed to maintain process integrity, steps must be taken to ensure continuous conformance to specifications. In certain circumstances, design of experiments may be required to determine the optimal environmental control parameters. Monitors should be strategically placed throughout the facility to assess current conditions and trigger an alarm or notification when deviations from defined standards occur. Don't overlook environmental conditions as a potentially significant contributor to process performance since parameters such as temperature, humidity, and technical cleanliness could impact process equipment and people as much as the quality of the product being produced.

Optimization Methods, Tools, and Techniques

There are several different methods that can be employed in the optimization of manufacturing processes and systems. They include process characterization, design of experiments, value stream design, and constraint management to name a few. Each of these techniques is part of a toolbox from which a tool can be selected to solve a specific problem or achieve a targeted objective. Let's consider several of these methods in the following pages of this chapter, starting with process characterization.

Process Characterization

The purpose of performing process characterization studies is to identify, quantify, and control significant sources of variation impacting production performance. The primary objective is to identify process parameters that significantly impact product quality and yield. This information can then be used to define the optimal operating ranges and acceptance criteria to realize a stable, consistent and predictable output. This approach to product and process optimization leverages gauge repeatability and reproducibility (GR&R) studies, capability (Cpk) studies, design of experiments and response surface analysis. The output of an effective process characterization study is a process control strategy that consists of the following:

- An understanding of significant sources of process variation.
- Measurable process characteristics and demonstrated measurement system capability.
- Line set-up parameters and nominal process settings.
- Process instructions and operating limits (max/min settings).
- Standard responses to deviations from process controls.

Process characterization studies should start early in product and process development and be completed prior to product design and manufacturing process validation.

Thoughts for Process Characterization

- Good process characteristics are measurable process characteristics.
- A sustainable process is a stable process that is fully characterized with appropriate process controls defined.
- Statistical data analysis plays a critical role in process characterization studies.
- A poorly understood and sub-optimized process can potentially result in a significant financial loss over a product's life cycle.
- A thoroughly characterized process, with the proper controls, can minimize process abnormalities and product defects.

Design of Experiments

A design of experiments (DoE) is a structured method conducted to understand the relationship between factors affecting a process and its resulting output. DoE is a good tool for helping understand the key factors contributing to significant process variation, factor interactions, and how these factors impact product and process output performance. The independent inputs (X) of a process are commonly referred to as factors. The dependent process outputs (Y) are considered responses. Factors can be materials or process settings such as time, temperature, pressure, etc. Every process output (response) exhibits known and unknown process variation, in addition to variation in the response variable measurement. DoE can help identify the few significant process inputs (or settings), from the trivial many, by combining several variables in one study. A properly executed DoE can help quantify process factors and test limits to establish process controls. Factorial DoE determines whether a relationship exists between critical factors and their degree of interaction. A Design of the Experiment can be performed as follows:

- Identify the key input factors impacting process performance.
- Determine the parameter limits (factor levels) to evaluate.
- Design an experiment.
- Run the experiment.
- Analyze the results.
- Determine next actions.

For more information, reference the many books and articles dedicated to this topic. Knowledge of statistics is helpful when looking to conduct an experimental design targeted at understanding the significant factors contributing to process output performance.

Response Surface Methods

Response surface methods (RSMs) complement a DoE by taking the factors identified as having the most significant impact on output results and using this information to determine the optimal (best) input parameter settings for process control. In essence, a response surface experiment is conducted to find the control factor levels that provide an optimal process response. This method, due to its statistical modeling approach, can only approximate the best parameter settings. Knowledge of statistics is recommended when using this methodology.

Process Flow Optimization Methods

The objective of process flow optimization is to streamline operations by achieving a steady and predictable production output at a rate consistent with customer demand. This can be realized by optimizing resource utilization while working to eliminate obstacles to flow, increasing productivity and product/service quality. Optimization requires an understanding of how processes interact with each other and the actions required to establish and maintain a steady, continuous flow of material and information between them. There are many ways to improve process flow, some of the more common methods include the application of value stream mapping (VSM) and design (VSD), constraint management, Spaghetti diagrams as well as the fundamental practice of employing the Plan-Do-Check-Act cycle for improvement. Let's briefly consider several of these applications.

Value Stream Mapping/Design

Process integration and optimization are about seamlessly connecting processes into a holistic system that can eventually be optimized for the movement of material and information throughout the supply chain. This holistic system, focused on a product family, is what we call the value stream. The "supply chain" refers to the interdependence of material and information flow from one process step or workplace to the next, creating product flow, like the movement of water in a stream. Supply refers to the delivery of material, components, and information required for production or product assembly needed at each process step while the "chain" denotes the connections between the sequential workstations required to complete a process. A value stream is all the actions (value creating and non-value creating) currently required to move a product from its raw material state to a final product for use or consumption.

Once a value stream has been identified, it can be mapped to visualize the activities and sequential flow of material and information within the system to establish a performance baseline and reference point for continuous improvement. When the current baseline has been defined, it's time to review it for optimization opportunities. Understanding the value stream requires knowing individual and team roles and responsibilities within the connected processes and encouraging team members to work together to shorten process time by eliminating waste. To do so, sub-processes and their deliverables must be highlighted and improvement opportunities identified and delegated to the lowest possible level of the organization for realization. Realizing these improvements often requires the application of structured methods for problem-solving and decision-making.

It's important to stress that VSM is not about the map, it's about understanding how material and information flows within operations. It helps to see beyond a single process and highlights obstacles to flow and sources of waste in the value stream while providing a vision or pathway for a future state. A completed current state map is the baseline reference for creating a future state design (e.g. map) that is complemented by an improvement plan intended to move an organization from the current to an optimized state of performance. The VSD map is the vision while the implementation plan is the blueprint for realizing that vision (e.g. future state). The book *Learning to See* by Rother & Shook is a great resource for understanding and applying this methodology. See Figure 4.4.1.

SIDEBAR: OPTIMIZATION BEYOND THE SHOP FLOOR

Manufacturing system management ensures operating processes remain stable and capable of generating consistent and desirable output results. Once established, process control becomes the building block for sustainable efficiency improvements. Global business dynamics requires continuous efficiency improvement of all enterprise operations to guarantee companies remain productive and competitive. To achieve excellence, the company mindset must move beyond optimization of discrete areas or work activities to optimize core business processes that span multiple departments and functions. At the highest levels of enterprise maturity, we must work as an integrated team, putting our individual and departmental interests aside in lieu of doing what's best and optimal for the enterprise. Companies that can achieve this level of coordination and cooperation are likely to survive turbulent times and thrive in their quest for enterprise excellence.

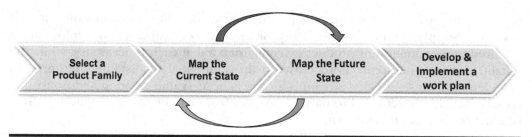

Figure 4.4.1 Value stream improvement.

Constraint Management

Theory of constraints (TOC) is a methodology for identifying the most significant limiting factor, constraint, or bottleneck in achieving a process target or business goal. The approach is to identify the bottleneck and systematically eliminate the factors contributing to the constraint, until it is no longer the gating item to achieving the target. The objective for manufacturing is to increase throughput with the aim of improving productivity and profitability. Once a constraint is eliminated, the focus then moves to the next constraint, always prioritizing the current bottleneck. This topic is covered in detail in the class book *The Goal: A Process of Ongoing Improvement* by Eliyahu M. Goldratt.

Spaghetti Diagram

The purpose of a Spaghetti diagram is to visualize the movement of people, materials, or information. A good Spaghetti diagram will clearly display the current flow patterns of the subject(s) being analyzed and reveal simplification opportunities. The diagram should document the current state of motion, based on observation. Upon completion, it can then be used as a reference or baseline for creating a more ideal or desired state of movement by rearranging equipment, eliminating steps, or changing the flow pattern. Preparing a Spaghetti diagram is a relatively simple process that involves the following steps:

- Identify an individual or team knowledgeable about the area or process to be improved or optimized.
- Prepare a motion diagram based on first-hand observation of the subject being studied. It's recommended that separate diagrams be used for people, material, or information flow, to avoid confusion.
- Review the resulting drawing(s) for problem areas such as excessive, inconsistent, or indirect movements. Consider moving equipment, eliminating work steps, or changing the starting point of activity to optimize the process.
- Create an ideal state of movement (e.g. a future-state flow map) and develop an action plan to transition from the current to desired state of flow.

Reference Figure 4.4.2 for a Spaghetti diagram example.

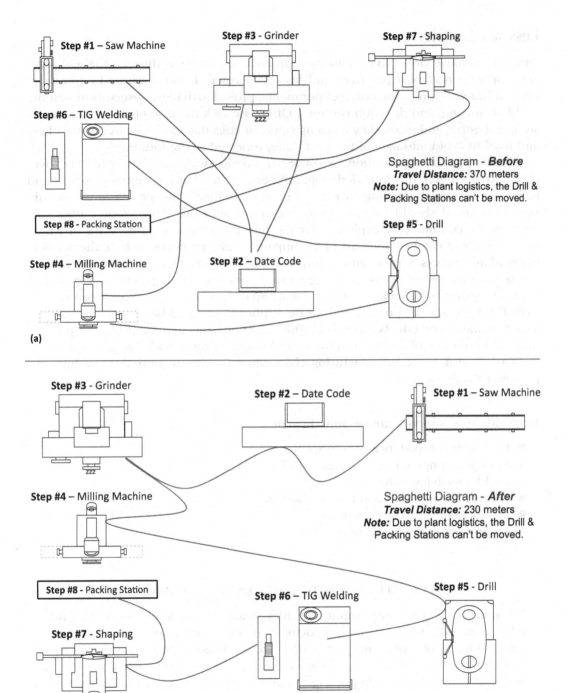

Figure 4.4.2 Spaghetti diagram example.

Lessons Learned

One way to move closer to optimizing a product or process is through the gathering and application of best practices and lessons learned. Lessons learned come from many different sources including experiments, project activities, observation, testing, problem-solving, and daily job routines. Often, we lack the discipline or patience to stop, recognize, and record key learning opportunities that can be shared with others and used to avoid mistakes of the past being repeated in the future.

Maturity, as an organization, is rooted in taking stock of what is learned over time and leveraging this knowledge and experience to change company culture and employee behaviors, leading to personal, organizational, and professional growth. Lessons learned should be seamlessly integrated into our daily work routines so they can be continuously exploited for the benefit of the company. Ways in which lessons learned can become part of a company's daily practices include the revising of standard processes, procedures, and work instructions. Updates to support documents such as control plans, parameter cards, product drawings, material specifications, equipment maintenance instructions, and operational guidelines help to retain "tribal" knowledge for present and future applications. In addition, documents such as failure mode and effects analysis (FMEA) and fault tree analysis (FTA) can provide invaluable records of historical lessons and engineering knowledge that can serve to assess and mitigate manufacturing risks when looking to optimize product and process robustness.

Tips: Valuable Sources for Lessons Learned

- Customer complaints and returns.
- Supplier component and material defects.
- Problem-solving outcomes.
- Manufacturing scrap and rework causes.
- Product failure analysis reports.
- Equipment downtime causes.

CASE STUDY: A PITCH FOR LESSON LEARNED

Manufacturing excellence cannot be achieved without the application of knowledge from lessons learned to drive focused and meaningful operational improvements. I knew of a plant in Texas with a cross-functional team that met once a week (every Tuesday at 10 am) for over 8 years to review every customer return and understand the issue impacting customer satisfaction. They worked on each issue to identify the root cause and implement corrective actions to prevent reoccurrence. Every action taken to prevent reoccurrence was confirmed effective and if not, the problem would be re-evaluated. This took a lot of work and discipline, but this persistence and commitment to quality and customer satisfaction paid off. Now, when the customer awards new business to the company, they insist the product be manufactured in the Texas plant.

Summary

Due to the changing dynamics of companies, customers, suppliers, and the world around us, optimization is a never-ending endeavor. Done right, optimization can result in a more flexible, lean, and responsive organization to changing customer demands. Optimization activities serve to improve operational efficiency by simplifying processes and procedures, standardizing on known best practices, capturing lessons learned, and focusing on process control parameters that matter. Simplicity and clarity help to reduce confusion and human error that often results in manufacturing inefficiencies.

Realistically speaking, optimization can never be fully achieved since the baseline is continuously shifting due to process variation, changing customer demands and technological advancements. However, what we can do is work to better understand the influential factors and how they impact operations while adapting our work environment to the changing conditions. The objective is to create a shop floor management system that can constantly adjust to a turbulent business environment.

As you may have come to realize, shop floor optimization is a full-time job. It's an activity that requires constant attention through process monitoring, control, and improvement. It's a perpetual cycle of plan-do-check-act in response to changing operating conditions. There is no silver bullet. It's about understanding what's required for operations, at any point in time, and striving to achieve the most efficient and cost-effective system possible. Optimization is unique for every facility and is influenced by the knowledge and actions of those responsible for realizing it. In the last chapter of this section, we will discuss the final frontier of daily shop floor management, that of manufacturing excellence.

Key Points

- The topic of optimization has a broad reach in manufacturing. It can include optimizing product quality, labor efficiency, equipment availability, material flow, or the entire value stream.
- Optimization is more likely to occur when employees share common goals and objectives.
- Process optimization is about improving operational efficiency.
- Optimization techniques can be applied to people, machines, methods, materials, and the environment.
- Improvements must be pursued with clear intent and justification.
- Practice putting ideas into application and learn from each experience.
- Optimizing system performance starts with the deployment of standards, procedures, and specifications that lead to process stability, capability, and control.
- Process control is a stepping stone for optimization which is a cornerstone of manufacturing excellence.
- Optimization results from the application of knowledge, best practices and lessons learned from across the organization. It's intentional and achieved through observation, experimentation, and continuous improvement.

◼ There are no limits to the opportunities one can pursue under the umbrella of optimization.
◼ To achieve excellence, the company mindset must move beyond the optimization of discrete work areas to the optimization of product value streams that span multiple departments and functions.

Sources

1. https://solutionsreview.com/business-process-management/the-secret-to-business-process-optimization/
2. Rother, M. & Shook, J. (2003). *Learning to see: Value stream mapping to add value and eliminate muda*. Lean enterprise institute.
3. Goldratt, E. M. (2016). *The goal: A process of ongoing improvement*. Routledge.

4.5

Manufacturing Excellence

Excellence is not a destination; it's a continuous journey that never ends.
~ Brian Tracy

Objective: Create a vision, mission, and strategy to realize manufacturing excellence and develop a plan to get there!

Overview

Manufacturing excellence is not a goal or an end state. There are no specific requirements to meet or rules to follow. Manufacturing excellence is a state of mind, an attitude, a behavior, and a desire for realizing the best one can be, at what one does, at all times. It means different things to different companies. It's defined by the values espoused by an organization and demonstrated through the behaviors, decisions, and actions of employees, at all levels of the company. Success is measured by performance results, customer satisfaction, competitive position, and organizational growth, among other factors used to measure a company's worth. In this chapter, we will explore some of the attributes associated with manufacturing excellence. Although the following is not an exhaustive list, it outlines some of the more common characteristics often associated with this topic:

- Enterprise alignment.
- Organizational structure.

DOI: 10.4324/b23307-27

- Operational discipline.
- Employee accountability.
- Agility mindset.
- Employee development.
- Process control.
- Continuous improvement.
- Technological advancement.
- Customer satisfaction.
- Supplier partnerships.
- Shop floor practices.

Let's briefly consider each of these topics in the following pages.

Enterprise Alignment

Manufacturing excellence is rooted in the vision, mission, and strategy an organization adopts and communicates to its people. The vision and mission must be relatable and articulate management expectations of employees. It also must be translated into strategic projects that are actionable and deployed throughout the organizational hierarchy. This helps provide a common purpose and aligns employees to work on meaningful activities that advance the company's ambitions. The vision, mission, and strategy must reinforce the desired behaviors, actions, and decisions the organization wants to promote in order to realize its pursuit of excellence. When all employees understand their purpose and how their actions serve that purpose, management can begin to lead the organization in the desired direction with clear resolve, purpose, and intent. In essence, the company vision and mission help define the mindset and behaviors expected of employees while management sets the examples of how to act and behave within the workplace. A lack of alignment can quickly derail an enterprise's ability to focus on what's important and achieve tangible success. *Enterprise alignment creates the organization direction, mindset, and behavior for success.*

Organizational Structure

Structure is an essential part of every organization, without which a degree of disorder reigns. In today's dynamic working environment, structure is more important than ever, since it's an effective countermeasure to increasing operational complexity. Structure brings order to complexity. It provides a framework within which to work and a direction where uncertainty once existed. It helps align employees toward a common way of performing their daily tasks while providing a guide for action and decision-making. It reduces process variation and facilitates process stability and control.

Structure also services as the foundation for manufacturing excellence. A carefully designed and executed work structure will lead the organization along the path to realizing its strategic objectives. An organizational structure is defined by its systems, processes, procedures, and work instructions, as well as the mindset of the people tasked to execute it. As with any structure, physical or otherwise, it must be maintained and continuously improved to weather the test of time as people work within it to advance the company's position in a turbulent, competitive, and often global marketplace. *Organizational structure outlines the path to success.*

Operational Discipline

Organizational structure must be complemented by operational discipline. Operational discipline is the practice of following rules, standards, and procedures. It's a code of behavior expected of employees to work in a measured way. In manufacturing, it takes discipline to follow the systems, methods, processes, and procedures required to produce predictable output that consistently satisfies customer expectations. A disciplined workforce is more likely to achieve operational excellence when a well-planned organizational structure is executed, monitored, controlled, and continuously improved.

Operational discipline must be an inherent part of organizational culture, along with the willingness to hold employees accountable and correct their behavior when not suitable for the work environment. Discipline must be demonstrated by leadership and made an expectation of all employees, while being reinforced through daily coaching and work-related interactions. Organizational structure is essentially meaningless without the operational discipline to maintain its integrity. Common ways operational discipline can be exercised and reinforced is by following operating standards and procedures, engaging in Gemba walks, practicing deviation management, conducting layered process audits, and holding employees accountable for executing their standard work routines. *Operational discipline ensures the path to success is followed.*

Employee Accountability

Accountability is reflected in an individual's willingness to accept responsibility for one's actions. In the corporate world, people's actions are judged by their ability to effectively exercise their roles and responsibilities as an employee and accept the consequences when they knowingly deviated from expectations. To ensure accountability, management must continuously take steps to address abnormalities in the actions or behaviors of employees perceived to deviate from organizational norms. Holding people accountable is simply the practice of "deviation management" applied to people, all for the sake of maintaining a stable and predictable operating environment. Manufacturing excellence requires a sound organizational structure and the operational discipline to reinforce and build upon the existing foundation. Employees contribute to this cause by understanding their roles and responsibilities, and continuously striving to realize their commitments, knowing their decisions, actions, and behaviors are a reflection of organizational excellence. *Holding employees accountable manages deviations from the path to success.*

SIDEBAR: ORGANIZE AND PRIORITIZE

All too often, we find ourselves overloaded with work, too much to do and not enough time or resources to do it. In response, some people get mad, fatigued, resentful, or even depressed. Responses to this type of stress vary widely among people with different personalities. One approach that appears to be relatively effective in dealing with this situation is to stop for a moment and think. Taking time to think gives you an opportunity to remove yourself from the current situation and consider it from a different perspective. It allows you to reflect on what's important, what really needs to get done and what will happen if certain tasks are not completed or completed on time. You may be surprised to find that things are not as critical or necessary to complete as you may have unwittingly assumed. So next time you start feeling the stress of a job or life's pressures, stop, take a breath, think, and prioritize your work. Knowing what to do next and what can wait a little longer may be all that's needed to take back control of your situation and move forward with confidence.

Agility Mindset

The ability to adapt quickly to today's changing business climate can be the difference between failure, survival, and profitability. The methods you employ and behaviors you display can provide a significant competitive advantage in the content, quality, and speed of goods and services delivered. Adopting a lean mindset requires a real behavioral change stemming from a significant emotional investment. Agility is the ability to move quickly in response to an uncertain and changing environment. It includes the ability to think, comprehend, and act rapidly with relative ease and confidence. Agile is as much a way of thinking as it is a way of working. Excellent organizations teach and coach people how to think and act in a logical, rational, and decisive way so valuable time and resources are not wasted deciding what to do next in abnormal or uncertain situations. *Agility provides the confidence and competence to continue along the path to success.*

Employee Development

Modern-day manufacturing must be flexible, innovative, lean, and adapt to changing customer demands in order to be and remain competitive. This requires a competent, capable, motivated, and empowered workforce with a unified vision, common goals, and objectives. To realize this expectation, employee development must become an inherent part of the company strategy. Resources must be allocated to ensure employees are continuously enhancing their knowledge and skills to efficiently maintain and improve operations while integrating technical advancements into the workplace that align with the company strategy.

Employee development can occur in many ways including training, mentoring, and coaching. A learn-by-doing approach, conducted through Kata projects, Kaizen events, and Jishuken workshops, can be an effective way to enhance people's understanding and capability in learning and applying new skills with the support of a facilitator and coworkers. Manufacturing excellence requires that companies strive to continuously improve all aspects of their operations. This includes processes, equipment, materials, the environment, and especially its people. A company is only as good as the people who work for it. Invest in your future by investing in people. *Success is possible because of the people who make it possible.*

Process Control

Process control is an essential discipline of shop floor operations and manufacturing excellence. An uncontrolled process is not predictable, and an unpredictable process can't be effectively managed. Getting "control" of a process starts with establishing standards from which deviations can be identified and defects eliminated through structured problem-solving. Process control can only be realized when process stability and capability are achieved. Maintaining stability and capability is the act of process control which relies on deviation management to highlight and permanently eliminate significant abnormalities. Process control is the foundation for continuous and sustainable productivity improvements and a steppingstone to system optimization and operational excellence. *Success is rooted in process control.*

Continuous Improvement

Standards help to reduce process variation, stabilizing systems, and establish a baseline for continuous improvement. Continuous improvement involves the elimination of process waste through the removal of non-value-added activities and starts to focus on system optimization by taking a holistic view of manufacturing operations from a product value stream perspective. When moving from process optimization to value stream optimization, we must give more attention to the connections that bring individual processes together and look to eliminate the obstacles that disrupt the flow of material and information between them, from raw material receipt to finished goods delivery, and sometimes beyond. A well-controlled operation is the building block for sustainable productivity improvements, process optimization, and manufacturing excellence. *Success is achieved through continuous improvements.*

SIDEBAR: WELCOME VISITORS

If a plant truly wants to learn, maintain, and improve their performance, let customers and other "outsiders" visit the plant. Every visit is an opportunity to solicit feedback and identify opportunities for improvement. Being subjected to the continuous scrutiny of outsiders will provide a diverse and subjective "fresh eyes" perspective of the plant and its performance. It will also require that plant personnel remain vigilant in their daily work routines in anticipation of the next visit. If not offered upfront, visitors should be asked directly for comments concerning their visit. Questions such as:

- What did you like about your visit to the plant; what impressed you the most?
- What opportunities for improvement did you observe during your visit?
- If you had to make one change about the plant, what change would you suggest?

Visitor feedback should be put on public display for all plant personnel to see. This can validate work currently being performed, highlight what's important to visitors and provide direction on what the plant needs to do next for continuous improvement. Manufacturing excellence is rooted in the openness of plants to accept feedback from key stakeholders and take action in pursuit of perfection.

Visits also provide a platform for plants to showcase their abilities and expertise. Transparency builds trust, trust facilitates comfort and comfort leads to long-term business relationships. So, next time someone wants to visit the plant, say yes, and open your doors to others in an effort to learn and improve. Accepting visits promotes good will and, under the right circumstances, creates advocates.

Technological Advancement

Technical advancement is dictated by continuous innovation to achieve leaner and more adaptable processes that are safer, cheaper, easier, and more efficient to deploy. These attributes are key for survival and competitive growth for many companies, especially those in high-cost countries. Enterprises that can rapidly integrate technological advances into their work environments will possess a distinct and significant advantage over their competitors if the technology deployed is proven to enhance productivity. Unfortunately, technology does not necessarily come cheap or easy. It must be strategic and focused on helping to achieve organizational goals and objectives with clear and proven benefits. Companies also require talented, motivated, and capable employees to scout, pilot, prove-out, and exploit the latest technologies, to suit their unique applications, in a meaningful, efficient and cost-effective manner. Optimization is achieved through persistence and experimentation while excellence is achieved through strategy deployment and continuous innovation. *Technology creates steppingstones to success.*

Customer Satisfaction

Customer satisfaction is the end goal since the customer is the reason why companies exist. Simply put, no customer, no money, no company. A product is made or a service is provided to a customer willing to pay for it, in order to satisfy a need. It's likely the need can be satisfied elsewhere, making it important to understand the wants, needs, and expectations of existing and potentially new customers. This can be achieved by building strong customer relationships and by knowing how best to serve customer needs. Part of building a customer rapport involves business transparency (within reason) and trust. Not being forthcoming, open, or honest can jeopardize a relationship. Words are cheap, so demonstrate your commitment to customer satisfaction through your actions. Listen, learn, understand, and then act. Remember, customers select suppliers they like, and ones that can solve their problems, while satisfying their needs. *Customer satisfaction is one of the measures for achieving success.*

Supplier Partnerships

Successful companies have good working relationships with their suppliers since they know their success relies on the quality and timely delivery of supplied materials and components. When companies work closely and cooperatively with their key suppliers, a strong, respectful, and long-term relationship often develops. This type of relationship helps in managing risk in material supply when economic downturns, forces of nature, and other unusual events cause supply chain disruptions. In fact, the realization of certain lean activities such as Just-in-Time inventory management extends beyond the walls of a manufacturing facility and depends on the competence of key suppliers to deliver a quality product on time, while maintaining a smooth process flow throughout the value stream. Ignoring the building of key supplier relationships can lead to indifference to business transactions and stifle the ability to achieve process optimization and manufacturing excellence. *Suppliers contribute to organizational success.*

Lessons Learned

Lessons learned can come from many different aspects of manufacturing including the customer, suppliers, new product development, equipment maintenance, and shop floor observations. Ignoring this valuable source of information can be detrimental to any organization not willing to learn from its missteps and mistakes. The biggest stumbling block to leveraging lessons learned is the fear of revealing weaknesses or mistakes to others. An organization that does not tolerate mistakes, will tend to hide them, creating an environment of fear and blame. Excellence in manufacturing encourages people to share their problems, concerns, and failings openly, without judgment or consequence. A proactive mindset translates organizational weaknesses into improvement opportunities and mistakes into waste for elimination. A learning organization is an improving organization and a continuously improving organization becomes a force for change. *Success is built by lessons learned.*

Shop Floor Practices

The shop floor has been the point of focus for this book. Many of the concepts discussed apply to other parts of the business as well. However, before we leave this chapter, I want to reiterate the importance of many practices already discussed since they are core to manufacturing excellence. The following are some of the essential shop floor practices that make manufacturing excellence possible. Consider them in light of your existing operations and as potential opportunities for improvement. Practices for realizing manufacturing excellence include:

- Defining operating standards and aggressive targets as performance measures.
- Establishing visual controls to highlight, monitor, and control process performance.
- Exchanging information through daily meetings and Gemba walks using real-time performance data.
- Prioritizing actions based on limited time and resources.
- Establishing standard work routines to articulate and prioritize employee activities.
- Engaging in deviation management as an essential activity of shop floor management.
- Stopping to permanently fix problems at the source.
- Taking time to go, see, and observe for more effective problem-solving and decision-making.
- Capturing and evaluating employee ideas for continuous improvement.
- Periodically verifying and validating compliance to operating practices, procedures, and work instructions.
- Maintaining equipment according to a thoughtfully prepared schedule.
- Engaging in structured problem-solving to permanently eliminate defects and other significant abnormalities.
- Mentoring, training, coaching, and developing employees wanting to remain relevant and continuously improve themselves.

In summary, manufacturing excellence reflects process operations rooted in process control and continuous improvement. Process Control is about maintaining a stable and capable process, so it consistently and reliably delivers results that meet customer requirements. If a company wants to grow and remain competitive, it must confirm that its core processes have a sound foundation upon which efficiency improvements are rooted. This foundation is established by ensuring stable and capable processes are in place along with the necessary discipline and accountability to maintain and sustain this desired state throughout a product's manufacturing life cycle. Once process stability, capability, and control are confirmed, the building blocks are now in place to drive a more efficient and sustainable operating environment for system optimization and manufacturing excellence. The goal is to deliver the safest, most efficient and cost-effective products and services that satisfy customer expectations while making a comfortable profit!

CASE STUDY: RULES OF ENGAGEMENT

This is a real story about a small manufacturing facility in Wisconsin that lost substantial sales when a major customer decided to divest from a business in which this manufacturer had a significant stake in supplying parts. The manufacturing team struggled to find replacement sales for about a year when their parent company announced the closing of a facility that produced a similar product line of recreational vehicle parts. The closure plan was to move the existing business to the Wisconsin location. This decision would breathe new life into this struggling facility, tripling its annual sales.

The transfer of business filled the facility to its manufacturing capacity, significantly increasing its workload. The leadership team was left to figure out how to manage the transfer of equipment and increase in material inventory with a shortage of people due to the low unemployment economy. The plant struggled to keep its existing workforce during the 9-month transfer process of moving 27 production lines to their location as significant facility preparations and operational disruptions overwhelmed the plant staff. The move was further complicated by pandemic-induced material shortages, a significant upgrade to their ERP system, and the unexpected announcement that the facility was being sold to a new owner only 2 months after the last line was transferred to the location. It was a perfect storm of events that threw the manufacturing team into a reactive mode of operation. They fought to stay on top of the many issues impacting daily shop floor activities, responding to immediate needs while leaving no time to address the fundamental problems they needed to eliminate for operational stability. They were plugging holes in the dike without time to patch them.

The team realized that their most immediate concern was material availability. Logistics was scheduling builds that ended prematurely due to part shortages causing line-down situations, operators without work, and delivery delays. In response to the instability, and their struggle to get out of their reactive mode of operation, the team decided to focus on two key objectives, 1) keep the high volume/high operator dependent lines running and 2) follow five rules of engagement to ensure material stability. The intent of the first objective was to keep people employed and allow product flowing out the door to generate revenue. The second objective was to establish "rules" that must be respected to ensure material availability on four high-volume lines over the next 30 days of production. These lines took priority over all others due to the value they generated.

These rules were required because the newly upgraded ERP system was showing more material available on site than actually existed. In turn, when scheduling production with this information, lines would often run out of material before completing the shift or production run, causing the line to stop, shorting the next delivery, and creating a dilemma of idle employees. This became a disruptive pattern of work that wreaked havoc on operations. Upon investigation, part of the problem became clear; employees were removing material from the warehouse without checking it out of the system. To combat this problem and reestablish manufacturing continuity, the following rules were implemented:

- Verify all needed parts are "on site" before scheduling a product build.
- Release the build schedule on the Friday before the build week (if not sooner).

- Only designated individuals can remove parts from the warehouse. All parts removed *must* be scanned out.
- Pack out and enter parts into the production system *at the end of each shift.*
- Scrap out parts in the system within 24 hours.

It was clearly communicated that if any of the rules could not be executed, the team must continue to work toward meeting them by understanding the obstacle preventing compliance and working to eliminate it by changing the system and/or our behavior so that compliance to all rules can be achieved without fail. Understanding root causes and implementing permanent corrective actions to eliminate obstacles impacting manufacturing stability is essential for achieving process control, optimization, and manufacturing excellence.

Key Points

- Enterprise alignment starts with a well-developed strategy that has been effectively deployed.
- Organizational structure is built on the standards, practices, and procedures used to run the company.
- Operational discipline and employee accountability make the organizational structure a reality.
- Enterprise agility is a reflection of company practices and employee behaviors.
- Strong supplier relationships facilitate the journey to manufacturing excellence.
- Excellence requires knowing your customers and making strategic adjustments accordingly.
- Future survival requires maintaining a competitive nature and being able to adapt the latest technology to the current manufacturing environment.
- Strategic failure is acceptable if it's used to learn and improve operations.
- Applying lessons learned can be one of the most effective ways to improve future performance.

NEXT GENERATION SHOP FLOOR

> *Our future success is directly proportional to our ability to understand, adopt and integrate new technology into our work.*
>
> ~ Sukant Ratnakar

Objective: Exploit technology and best practices for productivity improvement and competitive advantage.

Overview

All too often, organizations become complacent when operations are performing well, and the company is making a comfortable profit margin. Unfortunately, the manufacturing world does not stand still as competitors leverage disruptive technologies to drive productivity improvements that help them remain competitive. Thus, the "rules of the manufacturing game" are continuously changing, creating a dynamic, sometimes unpredictable but always challenging business environment. The winners of this game are those with the most efficient and agile operations that can quickly exploit new technologies and adapt to continually changing workplace dynamics and customer requirements.

Industry 4.0 Expectations

The new frontier of manufacturing has been defined by the term "Industry 4.0." This idea of Industry 4.0 refers to a new phase in the industrial revolution, focused on interconnectivity, automation, machine learning, and real-time data analysis for decision making. It's considered a "smart manufacturing" approach that leverages digital technologies in a production environment. The idea is to create a more holistic, interconnected supply chain system to manage machines, processes, and materials as a

DOI: 10.4324/b23307-28

collective. Interconnectivity of smart machines and sensors can be achieved through the internet of things (IoT), which connects the physical with the digital world, to enhance access and collaboration across departments, suppliers, equipment, locations, and people. The objective is to provide better operational control through immediate data access to increase productivity, improve processes, and drive business growth.

Enhanced communication between interconnected equipment, outfitted with monitors, actuators, and sensors, can extract real-time data for analysis to trigger preventive maintenance, automate controls, and optimize operational performance. Some of the key enablers behind the Industry 4.0 revolution include:

- **Big data** – management of large amounts of data which are compiled, stored, organized, and analyzed for patterns, trends, relationships, new opportunities, and other meaningful information.
- **Real-time data processing** – application of computer systems and machines to automatically and continuously process data to provide immediate information for review and necessary action.
- **Artificial intelligence** – computer learning which is used to perform tasks and make decisions without human intervention.
- **Machine to Machine (M2M)** – communication between two separate machines through wireless or wired networks.
- **Digitization** – collecting and converting different types of information into a digital format.
- **Cloud computing** – application of internet-based servers used to store, manage, and process information.
- **Machine learning** – application of artificial intelligence to facilitate machine learning that improves equipment performance, independent of human intervention.
- **Smart factory** – leveraging the above concepts in a coordinated way to create an ecosystem involving a complex network within manufacturing operations which could include inventory and planning, financials, customer relationships, supply chain management, and manufacturing execution. It's a highly digitized and connected production facility that employs concepts such as equipment sensing and machine learning to proactively manage operations and increase operating autonomy.

There are many different approaches to creating a "smarter" factory. For example, enhancing data visibility and transparency across the supply chain can accelerate product and service delivery without compromising quality or cost. A second example is the availability of real-time data analytics that predicts the need for equipment maintenance before unexpected failures occur. This would replace the time or routine based preventive maintenance approach that exists in many facilities today, leading to a more automated and streamlined system. Another example is the automation of asset tracking in the domain of logistics. Standard asset management tasks such as asset transfers, disposals, reclassifications, and adjustments can be streamlined and managed real time, centrally and remotely.

Propagation of the smart factory concept, throughout the product life cycle and supply chain, including design, sales, inventory, scheduling, quality, engineering, and customer support, can result in significant benefits to productivity and competitiveness. Predictive analytics, real-time data, internet-connected machinery, and automation can help the organization proactively solve maintenance and supply chain management issues. However, smart factories require management commitment and investment, complemented by a workforce capable of navigating the challenges inherent in the application of disruptive technologies. This also requires that you have the fundamentals in place to facilitate the integration of operational activities into a digital world.

Digital Transformation

When making the transformation to a more digital and interconnected working environment, consider how some of the lean manufacturing practices can be integrated into the new world order. The following are some examples:

- **Standard Work** – Standard work is the basis for successful implementation of Industry 4.0. Standardization drives process stability without which the application of technology becomes meaningless. Technology is built on a foundation of stable, capable, and controlled processes. Technology does not replace or compensate for poor manufacturing performance or practices, it enhances what already exists.
- **Visual Management** – The delivery of data and information will change with the application of digital technology. Manual displays will increasingly be replaced by digital images, with colors and sounds expected to play a significant role in attracting attention and triggering action. Key data and information are being delivered by computer tablets, smart watches, smart glasses, and smart phones, augmenting communication by providing real-time production status and information dashboards in different sizes, formats, and customizable to fit individual interests and needs. The use of digital technology, for visual data and information management, can enhance employee work performance in all areas of manufacturing and support functions.
- **Workplace Design** – Technology can be used to recognize employees and adjust a work area to individual needs by customizing their desk position, computer screen font size, or correspondence language, as examples. Technology-based workplace adjustments can also help employees with physical disabilities or health conditions perform their daily work routines.
- **Total Productive Maintenance** – Big data collection and data analytics can be used for predictive maintenance. Wearable devices can be employed for notification of equipment warnings and needed assistance. Smart glasses may also provide virtual work instructions for real-time application of preventive maintenance tasks.
- **Employee Engagement** – Workers will play an increasingly important role in the automation and digitalization age of Industry 4.0. They will be relied on to manage the supply chain, solve complex problems, make key decisions, and drive innovation forward with new, and complex systems.
- **Resource Consolidation** – Advances in technology has facilitated remote work, making it possible for some employees to "work from anywhere". This opens

the opportunity for consolidation of certain work functions or activities to a central (lower cost) location, reducing the need for redundant activities within a facility or between multiple facilities.

◼ **Simplification** – One way to reduce the cycle time of a product or process is through simplification. As processes become more complex, we must focus on simplifying the understanding of how they work and the need to be managed for efficiency and success. This can be achieved when systems can intelligently interact with each other without human intervention.

Productivity Improvements That Make a Difference

Few would dispute that the future of manufacturing is changing and, in some cases, changing quickly. In order to stay ahead of the competition, companies need to innovate. This not only applies to products, but also the processes used to produce products. Speed and agility are needed to adapt to the dynamics of manufacturing. Continuous cost reductions must also be pursued to remain efficient. A typical and immediate response to reducing costs is through labor reduction. However, this must be done strategically to avoid a negative impact on product quality and employee morale. It's better to take a holistic approach to productivity improvement by considering all opportunities, including equipment reliability, material savings, process simplification, and environmental impacts.

Many facilities are taking a smart approach to productivity gains through the application of technology. Technology often requires organizational knowledge and a highly skilled workforce who can visualize, adapt, and integrate the latest hardware and software enhancements into an existing manufacturing environment with minimal impact to operations. More importantly, technology needs to be justified. Simply pursuing technological advancement without due consideration to the benefits can be a disaster in the making. Understanding what can be done and assessing potential benefits, by properly vetting new opportunities, can go a long way to avoiding the pitfalls of implementing technology that does not deliver on its promises.

There are many opportunities to consider when it comes to cost-saving and efficiency improvement technologies. In the following pages, we will touch on some of those opportunities already in practice and others that have the potential to provide real value when considered in the context of existing needs, opportunities for improvement, and revolutionary advancements. I will present various applications of technology and lean concepts for your consideration: some proven while others remain in development, as they continue to evolve over time. Let's start the review with the growing application of collaborative robots (Cobots) in the workforce.

Part 1: Technical Innovations

Collaborative Robots (Cobots)

Collaborative robots or "cobots", as they are often called, are computer-controlled robots designed to assist people in the workplace. They often work side-by-side with production line operators. They are good at performing repetitive work tasks,

Figure 5.0.1   Cobot.

such as pick and place, freeing operators to perform more complex work requiring flexibility and observational skills. The difference between an industrial robot and cobot (e.g. service robot) is that an industrial robot is designed to work independently of people, in a separate, protected area or location to prevent human contact. A cobot can work without the barriers of an industrial robot, in parallel or in concert with people. Cobots are equipped with sensors to detect the presence of people within their workspace, allowing them to slow down or stop without the threat of human injury. They can be very cost effective when considering they can work 24/7, don't stop to rest, eat, or use the toilet. They don't expect pay or benefits, won't get "injured" on the job, and will not walk out on strike. They will do dull, dirty, and dangerous jobs that employees avoid, and tend to be highly reliable, working for years with minimal downtime for scheduled maintenance. They can be a plant manager's best employee!

Due to their relative ease of installation and programming, cobots can be quickly integrated into a production environment. Skilled technicians are required for installation and scheduled maintenance. Their capability and capacity for handling diverse work requirements and increasing loads is improving with time as designers consider additional applications and feedback of avid users. Cobots are the workers of the future, especially in high-cost countries where labor significantly contributes to product costs. See Figure 5.0.1 for a cobot picture.

Robotic Process Automation (RPA)

Robotic process automation, or RPA for short, is a technology that enables software 'robots' to carry out repetitive, rule-based digital tasks. Humans commonly perform these tasks through the user interface, via a mouse and keyboard. RPA robots can mimic these human actions quicker, more accurately and consistently. Application of RPA can improve operational efficiency when manual, highly repetitive, low exception rate, and standard electronic readable inputs are combined with a rule-based process. Robotic process automation can streamline workflows, making operations more efficient, flexible, and responsive. Software robots (or bots) can also boost employee moral by eliminating the highly repetitive, mundane tasks from people's work routines.

Automated Guided Vehicles (AGVs)

Automated guided vehicles, AGVs for short, are computer-controlled wheeled vehicles that carry or transport material throughout the facility, independent of an operation or driver. Software, in combination with a sensors-based guidance system, is used to direct a vehicle through the facility to designated locations. AGVs can work among employees, sensing people's proximity to the vehicle and adjust their movements by slowing down, stopping, or redirecting their motion in order to avoid "obstacles" in their path. Much like Cobots, they don't need to rest, eat, or go to the toilet. However, they do need to be periodically charged since they run on battery. When not in use, they are programmed to return to a base station where they can be charged while waiting for their next task.

There are many types of AGVs, including ones designed for transport, lifting, and towing. Basic units, such as an automated guided cart (AGC), can have a simple navigation system consisting of a magnetic tape placed along its travel path. More sophisticated sensor-based vehicles use artificial intelligence to pilot their movements. Hospitals use these types of units to transport items such as meals, linens, biohazards, and surgical supplies. Forklift AGVs can replace a human-operated unit. Weight barring units can tow, load, and unload material at different locations, freeing material handlers to do other tasks. These driverless train-like units typically follow designated paths along the shop floor. Load handler units can carry discrete objects such as a single item, pallets, or totes containing multiple items. There are units designed for heavier objects with self-loading capability and more technically advanced units called autonomous mobile robots (AMRs) that use intelligent navigation systems connected to sensors and cameras that guide them around obstacles such as people and immobile objects.

Micro AGVs

Micro AGVs are entering the manufacturing scene to transport work-in-process materials to various bench-top assembly stations. Proximity sensors are used to relay instructions to an on-board microcontroller that directs the unit to specific assembly locations within the workstation. This is a relatively new technology, in an early stage of development, which is likely to evolve quickly over the coming years.

Smart Glasses

Smart glasses are considered wearable computer glasses that can run self-contained mobile applications and communicate with the internet via voice commands. They can capture hands-free digital images, video, or audio recordings which can be shared real time with others. Smart glasses can also be set up to access information to complete a task, alert users, and to communication data such as calls, text messages, and emails. They can facilitate remote collaboration in quick response to a problem when an expert is not immediately available on site to assist. Organizations are also looking into using them to create digital work instructions that can be used by new employees for training or to perform unfamiliar jobs. There will be more to come on this topic as manufacturers find new and different ways to exploit this technology.

Centralized (Automated) Inspection (AOI)

Centralizing inspection activities to a single area or "hub" within a manufacturing facility can significantly increase operational efficiency by reducing the number of inspectors needed when multiple production lines exist. Many companies are pursuing productivity improvement through centralized inspection where work once conducted at the production line is now being performed at a central location. This can result in a reduction of human resources that are focused at one location performing inspections for multiple production lines. This tight-net group of individuals can work in concert to disposition parts coming from many lines needing inspection services.

There may be some logistics required to move parts in and out of the inspection area, but the cost of doing so, over the long run, may be insignificant when compared to the labor costs required to maintain individual inspection stations. Various factors must be considered when determining the potential benefits of this approach such as types of material to be inspected, process flow, inspection requirements, test equipment needed, and environmental controls to perform the work. When equipment is centralized, product must be moved from the production line to the central location for inspection. This can add costs in the form of conveyors or disruption of production from its normal flow to the inspection area. However, a reduction in inspection equipment and skilled manpower to a few experts, working cooperatively together, can provide a significant savings to manufacturing, when centralized. This approach may also reduce wait time for product to be inspected, for inspectors to arrive, or the need for an inspector to move continuously from one inspection station to another. A centralized function also becomes more independent of the people and equipment for which the inspection is being done. Often inspections can be done remotely from where parts are located, using digital imagery. This can eliminate the need for part transport but may require parts to be controlled prior to disposition.

Unfortunately, central inspection may be limited by product size, weight and volume constraints, frequency of inspections, and the continuous movement of product from one operation to another, while maintaining a steady flow. Centralized inspection may also increase the risk of accumulating defective parts, if part inspection is not performed in a timely manner. This can occur when defects are generated in a high-volume environment, overwhelming inspectors. However, this can be managed through Jidoka practices where the process is stopped when multiple defects are identified within a short time period. It's important to remember that inspection costs money and is a form of waste. The ultimate goal is to eliminate inspection by ensuring quality at the source.

CASE STUDY: VISUAL INSPECTION WITH ARTIFICIAL INTELLIGENCE

I observed the result of a manufacturing team that used artificial intelligence (AI) technology to replace an operator-based inspection system where digital images are now used to accept and reject product with minimal human intervention. Inspection decisions can be impacted by image lighting and process variations causing potentially inaccurate decisions. To achieve more consistent and accurate inspection results, AI was employed to learn the characteristics of good and bad parts based on a multitude of parts already classified as good and bad by experienced inspectors. The AI-enhanced inspection equipment becomes more accurate or "intelligent" as to what's acceptable and unacceptable the more parts it characterized.

Months of randomly classified images, good (OK) and bad (Not OK or NOK), were fed into a convolutional neural network (CNN) to develop, test, and validate the model. Upon model validation, the "intelligent" inspection system was released in a monitor mode, comparing inspector decisions with model decisions. Results were reviewed and the model refined to achieve optimal and repeatable results. Successful application of the machine learning algorithm with automated optical inspection (AOI) equipment occurred when inspection accuracy was demonstrated to be better than a human inspector for 3 months. Not only did inspection accuracy improve, inspection cycle time (per part) decreased when human intervention was essentially eliminated in the disposition of part quality.

Fixed Process Cell

Many traditional electronics manufacturing assembly cells can be characterized as "complexity in motion" with motion of the assembly equipment being a root cause of many equipment failures over time. There was an initiative taken by an innovative manufacturing plant in Texas to rethink the assembly cell concept by reducing equipment motion during assembly. To do so, they challenged the typical approach to cell design and asked the question "What if the *product* moved instead of the process?" The objective was to minimize complexity, equipment motion, and waste by reducing the number of moving parts within the cell. The result was movement of the product within a cell of stationary assembly equipment. For example, when dispensing a coating or potting gel, the product moved, not the dispensing gun. When screwing, the product moved to the correct position, under the screw gun, which was in a fixed position within the cell. This approach, where applicable, reduced equipment downtime, capital, and maintenance costs. It also reclaimed floorspace since the manufacturing cell footprint was often reduced. This concept is only applicable under certain conditions where excessive equipment movement is the primary contributor to unplanned and unacceptable downtime.

Autonomous (Digital) Workforce

An autonomous (digital) workforce leverages the unique and unifying attributes of intelligent automation concepts such as robotic process automation (RPA), artificial intelligence, and machine learning to reduce the burden of business process work on employees, freeing up workforce capacity while ensuring automated tasks are completed with speed, efficiency, and accuracy. The primary objective is to automate the manual, stable, repetitive, and mundane workforce tasks. Creating an autonomous digital workforce requires the automated execution of business processes using software to emulate human interaction with targeted applications and systems, while exploiting the use of AI and cognitive data processing to mimic how the human mind reads structured documents. This includes simulation of the human intellect of learning, reasoning, and self-correction to manage unstructured interactions such as emails and short message service (SMS) communications.

Part 2: Thinking Differently

Big Data

Data sets have grown significantly over time becoming larger and more complex to deal with via traditional data processing software applications. As data grows, so does the challenge of gathering, storing, analyzing, sharing, and visualizing it. Big data, from today's perspective, is characterized by larger volumes and varieties (types) of data, arriving at accelerated rates. The objective is to extract data from equipment and store it in a database for analysis, system messaging, and action. The idea behind big data in manufacturing is to collect machine data, analyze it for meaningful information, and use what's learned from data correlations and trends for predictive and preventive purposes. For example, predictive maintenance can help avoid unplanned equipment downtime or prevent defects through cause and effect analysis. The benefits of big data analysis are vast and will evolve over time as we become more efficient at extracting valuable data and analyzing it for significant information that leads to sustainable action.

SIDEBAR: IMPROVEMENT PROJECTS USING AI

There are many ways to benefit from the application of artificial intelligence (AI) in manufacturing. One automotive parts supplier took the initiative to promote the use of artificial intelligence by enabling easier and more user-friendly access to data by pre-processing and co-locating data from a data lake into a flat table. This approach eliminated the need for users to spend time on data preparation and allowed them to focus their efforts on analyzing and understanding data for exploitation. As an example, one team used data from the flat table file to pursue predictive quality in the areas of solder paste inspection. The inspection machine measures thousands of part soldering characteristics looking for anomalies (deviations from specifications). When an anomaly is detected, the part is detained to avoid non-conforming product from escaping to customers.

Carbon-neutrality has become more important and awareness of energy consumption more prevalent over the years. In response, this same manufacturer initiated a second project to develop an energy management platform for transparent monitoring of equipment energy consumption and tracking usage anomalies. This approach allowed the factory to review consumption data and take action to improve its energy efficiency by reducing wasteful usage in high consumption areas of the plant.

Predictive maintenance is another area of manufacturing that can exploit the benefits of artificial intelligence. Predictive maintenance uses data analysis tools to detect operational anomalies and potential defects in equipment so they can be highlighted and addressed before equipment failure occurs. As one example, digital input/output sensors can collect data (such as voltage, pressure, current, temperature, or position) to monitor the current state of operating equipment and continually compare the data to a reference or standard value. Significant changes from nominal value can then trigger a warning, leading to preventive maintenance that helps avoid unexpected machine damage or significant downtime.

Artificial intelligence is also being used in manufacturing vision systems that do repetitive tasks using neural networks. In this case, vision systems are used to detect morphological defects on different surface materials and part shapes. Abnormalities detected are then isolated for further investigation. These are just several examples of how AI is used to advance manufacturing knowledge and provide insight on how to leverage real-time data for productivity improvements.

Equipment Robustness (Reliability and Durability) – A Case for Data Analytics

The reliability and durability of similar equipment can differ significantly, depending on the environment and application in which it's being used. For example, a car's preventive maintenance routines can be based on mileage, time, or driving conditions. Mileage can cause vehicle components to wear, time can cause those same parts to age while the driving environment can accelerate both wear and aging over a shorter time period. Similar to developing a vehicle maintenance schedule, the reliability and durability of manufacturing equipment can be influenced by knowledge gained through observation and experience during regular maintenance routines and repairs. Lessons learned from equipment maintenance and repair reflect the conditions under which the equipment is used. Not spending time to document observations, problems encountered, and corrective actions taken results in the loss of valuable information. If you collect and analyze the causes for equipment downtime and the actions taken to avoid or prevent these issues in the future, you can leverage this knowledge when looking to design or purchase new equipment. Use insights acquired from lessons learned to provide equipment manufacturers with updated material and component specifications that reflect your equipment usage conditions to create a more robust machine. In effect, you can optimize the next generation of equipment, for its intended application, by exploiting and sharing knowledge acquired from historical maintenance records.

If the original equipment manufacturer (OEM) is unwilling to make requested changes based on lessons learned from your unique user environment, consider another supplier or making the upgrades yourself, when the equipment arrives on the dock, before it's installed and qualified for use in production. A cost–benefit analysis should be done to justify any potential changes to existing equipment already installed in the facility. Proactively speaking, consider the cost of making value-added changes or modifications by the OEM before purchasing new equipment relative to the cost of equipment upgrades, done on site, prior to the start of production. This approach requires that reliable data and information be used in the cost–benefit analysis. Use your knowledge and experience for competitive advantage since every improvement counts!

Predictive Equipment Maintenance

Predictive equipment maintenance is an approach for evaluating the actual condition of in-service equipment to estimate when to perform maintenance. This method replaces a time-based preventive maintenance program by performing maintenance when needed, leading to cost savings. This condition-based maintenance strategy looks for degradations in equipment parameters or performance as a trigger for action. Additional benefits of predictive maintenance are convenient scheduling for corrective maintenance and avoidance of downtime due to unexpected equipment failures. Continuous data collection, monitoring, and analysis are key for detecting signs of equipment faults or degradation. The availability of critical spare parts or ability to make necessary parts on demand (e.g. machining, 3D printing) can significantly contribute to equipment availability.

Predictive differs from preventive maintenance in that predictive maintenance is in response to fault detection in a piece of equipment associated with a cause and effective relationship over time, whereas preventive maintenance relies on average or expected life statistics. A predictive approach is considered the next step in optimizing equipment maintenance by facilitating early detection of a problem and empowering employees to investigate and solve potential problems before they become bigger issues. Predictive analytics changes the questions from "what has happened?" or "why did it happen?" to "what is going to happen?" and "what can we do to prevent it from happening?" Predictive analytics can shift manufacturer's engagement with equipment from preventive to predictive maintenance. See Figure 5.0.2 for the evolution to predictive maintenance.

Spare Parts Management

Spare parts inventory costs money. Parts must be purchased and stored when not currently needed. However, the lack of spare parts, especially long-lead time parts, can significantly impact production by extending equipment downtime. This is why spare parts management is a strategic activity of lean organizations. Consideration must be given to critical and custom components, component availability, shelf

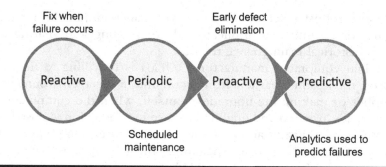

Figure 5.0.2 **Evolution of Predictive Maintenance.**

life, component delivery lead times, and the ability to make spare parts in-house. Machining and 3D printing spare parts, within the facility, can greatly reduce the risk of equipment availability delays due to waiting on long-lead time parts for equipment repair. Assessing the ability to make parts in-house and preparing to do so can significantly minimize equipment downtime and enhance productivity. Preparing a strategy for managing critical spare parts is another step toward manufacturing excellence.

3D Printing of Spare Parts

Three-dimensional printing uses computer controls to create a three-dimensional object, layer by layer, with a combination of materials, including plastics, liquids, and powder grains. These materials are fused together to produce tangible objects from a computer-aided design (CAD) or digital 3D model. As the flexibility and diversity of 3D printing has improved over time, so has the ability to make more complex and intricate objects. This capability has led many plants to reconsider the management of spare parts in their operations. They have come to realize that many of their spare parts stock can be made with in-house 3D printers, permitting a reduction in inventory and the corresponding cost and storage space associated with it. Consideration must be given to what critical parts could and should be made in-house versus what is more appropriate and practical to purchase, when it's needed. A cost–benefit analysis may be required when targeting parts for in-house 3D printing.

Suitcase Tester (or Other On-Site Testing Devices)

Suitcase testers or portable test units, provided to customers, from suppliers, can significantly reduce the number of warranty returns a supplier receives. A portable tester will allow the customer to test a "questionable" part for functionality before deciding to return the part to its supplier for analysis and disposition. This approach also provides customers with the ability to quickly determine if a questionable part is a likely contributor (or not) to a problem or concern under investigation. The cost associated with developing, building, and distributing testers should be weighed against the benefits of reduced warranty returns, and the efforts required to analyze and disposition returned units.

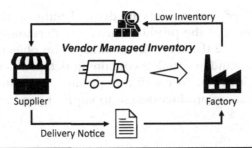

Figure 5.0.3   Vendor Managed Inventory.

Vendor Managed Inventory

Vendor managed inventory (VMI) is an inventory management and order fulfillment model where the buyer provides information to the vendor who takes full responsibility for maintaining the buyer's material inventory for products ordered, usually located at the buyer's consumption facility. This business model helps avoid out-of-stock issues and reduces supply chain inventory. It can be used to regulate or reduce plant inventory and corresponding material carrying costs.

I observed vending machines strategically located in a manufacturing plant containing such items as gloves, electrostatic discharge (ESD) foot straps, cleaning supplies, and other materials used by factory floor personnel. These are typically consumable products that require periodic replenishment. Workers scanned their company ID badge at the vending machine to obtain items they needed for work. The supplier is paid once an item is extracted for use and is usually responsible for replenishing stock. Plants may want to consider the potential benefits of pursuing VMI for their locations to reduce inventories and complexity of logistics (Figure 5.0.3).

Machine Learning

Machine learning is a form of artificial intelligence where computer algorithms are continuously improved through data processing and experience. Machine learning algorithms start with a model based on existing data and build upon it to make independent predictions and decisions that become better with increasing data over time. The idea behind machine learning is to collect machine data, analyze it, and use what's learned from data correlations and trends for predictive and preventive actions. Machine algorithms can also use learned information to perform specific tasks such as medicine, speech and facial recognition, assessing product quality, and detecting product defects using vision systems.

Modular Product Architecture

Modular product architecture is a practice used in product design that applies the principles of modularity. It involves an assembly of modules with unique functions and strategies. Modular product architecture consists of interchangeable building blocks (modules) that can be arranged into a series of product variants. To adopt this architectural approach, the product portfolio is decomposed into reusable elements, requirements are clustered, and variants are defined within a module. These activities are followed by alignment of the product lines by understanding the future product roadmap in order to derive the modular roadmap. The objective is to create transparency and provide a standard catalog containing detailed descriptions of reusable elements to enable their flexible use in future products consistent with the roadmap. This approach should allow product design to support product line and equipment reuse in manufacturing.

Mobile Devices

Mobile devices such as cell phones, smart watches, and tablets are no longer for personel use alone, they are being extensively used in business, and especially in manufacturing, to access emails, chat with colleagues, receive notifications, access real-time data, turn equipment on and off, and much more. These devices are reshaping the future workplace. However, security will likely be a primary concern since remote access to company data and information opens the door for others to access this information, sometimes for nefarious purposes. Regardless, this is an open space that continues to evolve and will likely do so with time.

Gamification

Gaming has played a significant role in the younger generation's leisure time, especially in how they interact with the world. As this new workforce enters manufacturing, there has been increased interest in incorporating the concept of gamification into work activities. Gamification is the application of game theory and design elements to non-game scenarios. Gaming in the manufacturing realm looks to make highly repetitive and tightly controlled work more engaging and links output achievement with recognition and reward to boost employee motivation and moral.

Gamification can help to incentivize employees, encourage healthy competition, and reward high performers within the work environment. When first starting with gamification, keep it simple and make the rules clear. Consider what motivates people and implement games that excite and inspire them. Celebrate the winners and reward the players. Simple rewards such as points, badges, streaks, challenges, titles, and leaderboards are likely to keep participants motivated. Highlight outstanding performance to headline your best performers. Regardless of what you do as a company, it's important to recognize that gaming is not for everyone. Respect those who may not be interested in game participation as part of their work activities.

Environmental Sustainability

Sustainability is the ability to maintain something such as a process, system, or the environment, at a certain rate or level. In manufacturing, it's more concerned with using natural resources respectfully to avoid any significant harm to people or nature's ecological balance. It's about meeting business needs without compromising the ability of future generations to satisfy their own needs. In doing so, there is a push to pursue more renewable and clean energy sources such as solar, wind, geothermal, and hydro energy.

There are ways for companies to boost their contribution to a more sustainable workplace, including the implementation of recycle programs, conserving energy, going paperless, reuse of equipment and materials, supporting green suppliers, and investing in sustainable transportation. These are not necessarily cheap or easy to pursue; however, the consequence of not doing so may have a significant impact on company perception, future business opportunities, and more importantly, global climate change.

Consciousness of a company's carbon footprint must also be taken into consideration. Carbon footprint refers to the amount of greenhouse and other gases released into the atmosphere, including carbon dioxide, methane, nitrous oxide, and chlorofluorocarbons. It's a measure of one's impact on the climate and climate change. These factors and others must be considered when managing the next-generation shop floor.

Key Points

- Invest in emerging technologies wisely. Take time to evaluate their benefits and impact on bottom-line manufacturing costs.
- Use the company business strategy to prepare manufacturing for the future.
- Application of Industry 4.0 concepts will attract young talent and help retain talent over time.
- Big Data, combined with the IoT, has opened a new era of productivity improvement possibilities.
- Predictive maintenance differs from preventive maintenance since it relies on the actual condition of equipment, rather than average or expected life statistics, to predict when maintenance will be required.
- Three-dimensional printing spare parts in-house can reduce costs and time for equipment repair.
- Mobile devices are enhancing workforce flexibility and the ability to work from anywhere in the world.
- Gamification can help engage employees to work in a highly standardized and repetitive work environment.
- Environmental sustainability must be part of a company's long-term strategy for success.

Sources

https://6river.com/what-are-automated-guided-vehicles/
https://www.epicor.com/en/resources/articles/what-is-industry-4-0/

Appendix 1 – Daily Shop Floor Management Checklist

This shop floor management checklist is a guide that presents basic shop floor activities. It's not comprehensive, but it includes many of the fundamental activities needed for continuous and effective shop floor management.

General

- **Company vision, mission, and strategy** have been deployed and are understood.
- **Strategic projects** are being executed, monitored, and controlled.
- **Design tools** are being deployed (Design for Manufacturing [DfM], Design for Six Sigma [DFSS], Design of Experiments [DoE], etc.)

Production

- The **build schedule** is available for this week and next week.
- No known obstacles are preventing the **build schedule** from being completed this week and next week.
- A daily **ship schedule** has been prepared for the week.
- Needed people, materials, equipment, and procedures are **available to build** for the week.

Manpower/People

- Employee **training, cross-training, and development** are on schedule.
- **Standard work instructions** are being respected.
- Operators are maintaining their defined **workflow patterns**.
- **Ergonomic practices** have been considered and are being implemented.

Materials

- Materials are received in **compliance** with their specifications.
- Material **replenishment** is happening as planned/on schedule.
- Material **handling procedures** are defined and respected.

- Material **tracking system** is periodically confirmed as effective.
- Periodic **cycle counts** are being performed by comparing actual material to inventory records.

Methods/Process

- **5S** cleanliness and orderliness standards are being monitored/maintained.
- **Work instructions** have been defined, standardized, made available, are effective, and followed.
- **Visual controls** are available, up-to-date, and being monitored for patterns, trends, and deviations.
- **Deviation management** is being effectively executed.
- **Problem-solving** activities are on-going.
- **Scrap** is being reviewed, dispositioned, and top issues addressed.
- **Customer returns** are being reviewed and used to improve performance.
- **Standard work routines** are documented and being effectively deployed.
- **Gemba walks** are occurring with good participation.
- **Action items** are being recorded, prioritized, and closed in a timely manner.
- **Process control documents** such as flow diagrams, process failure mode and effects analysis (PFMEA), control plans, and parameter cards are available, understood, and respected.
- **Performance reviews** are being held to review key process indicator (KPI) trends and follow - up on systemic actions.
- **Communication/escalation** meetings are being held from line leaders to senior management.
- **Statistical process control** is in place with out of control points understood and addressed.
- **Jidoka** has been implemented and is effectively being deployed.
- **Poka-Yokes** (error proofing) are in place, working, and effective.
- **Electrostatic discharge** (ESD) protections are in place for static electricity-sensitive materials.
- **Changeovers** are being monitored, controlled, and targeted for cycle time reduction.
- **Layered process audits** are defined, executed, and generating result for process improvement.

Visual Controls (Data Display for Process Identification and Monitoring)

- **Signage** for production lines and equipment is in place.
- **Key process indicators** (KPIs) and process targets are defined and posted.
- **KPI trends** are monitored for patterns and deviations from target.
- **Data collection** formats and templates are prepared and available.
- **Area/line information boards** are available and up-to-date.

Machine/Equipment/Tooling/Fixtures/Gauges

- **Calibration** is being performed per schedule.
- **Maintenance** (TPM) is being performed on schedule.
- **Improvements** are being made to equipment and equipment maintenance activities, based on lessons learned.
- A **tracking system** is in place to quickly locate needed equipment, tooling, fixtures, and gauges on demand.

Facility

- **Technical cleanliness** checks are being performed on schedule.
- **Climate controls** (temperature and humidity) are being maintained within required tolerances.
- **Safety** policy and procedures are being followed.
- **Environmental** practices are in place and are being respected.

Continuous Improvement

- **Changes** are being properly managed and confirmed as sustainable.
- **Employee improvement ideas** are being captured and properly managed.
- **Kaizen events and Jishuken workshops** are being performed and driving improvements.

Appendix 2 – Manufacturing Operations Manual – *Example*

The following is a simple example of a manufacturing operations manual. It's to be used for information only (Figure A2.1).

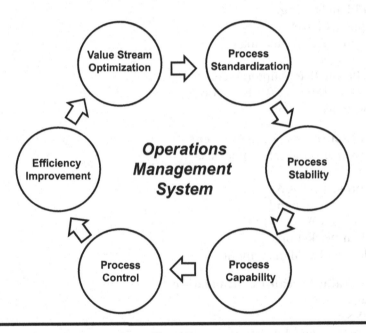

Figure A2.1   Manufacturing Operations Life Cycle.

Table of Contents:
Purpose
Six Phases of Process Maturity

1) **Process Standardization**
   - Standard work instructions.
   - Standard operation sheet.
   - Work combination table and cycle time diagram.
   - 5S method.

2) **Process Stability**
   ■ Production line output chart.
   ■ Statistical process control charts.
   ■ Run charts.
   ■ Key process indicators.
   ■ Jidoka.

3) **Process Capability**
   ■ KPI performance charts with targets.
   ■ Performance trends.

4) **Process Control**
   ■ Standard work routines.
   ■ Layered process audits.
   ■ Visual Controls/Line Information Board.
   ■ Deviation Management.
      – Problem-Solving.
      – Paynter Charts.
   ■ Daily Production Meetings.
   ■ Gemba walks.
   ■ Total Productive Maintenance.
   ■ Poka-Yoke (Mistake/Error) Proofing.
   ■ Performance Reviews.

5) **Process Continuous Improvement**
   ■ Process mapping/value analysis.
   ■ Kaizen events.
   ■ Kata projects/PDCA.
   ■ Change Management.
   ■ Self-directed Work Teams.
   ■ Jishuken Workshops.
   ■ Employee Idea Management.

6) **Process (Value Stream) Optimization**
   ■ Pull and Flow.
   ■ VSM/VSD.
   ■ Vendor Managed Inventory.
   ■ Kanban.
   ■ Heijunka.
   ■ Supermarkets/First In First Out.
   ■ Strategic Projects.

Purpose

This document establishes a framework for the **[Name]** manufacturing plants to build a production system that achieves, maintains, and continuously improves operations. Each plant will prepare an operating system that meets the intent of this standard while taking into consideration safety, quality, productivity, and cost.

Place an example of Manufacturing Operations Life Cycle here!

The **operations system** establishes the vision and direction for achieving a more operationally efficient enterprise. The phases are presented in a sequence that follows the maturity of a production operation, starting with process standardization, and leads to the optimization of product family value streams. Most facilities will exhibit various phases of maturity, typically reflected in the age of a manufacturing line and integrity of line design. What's important is to recognize a production line's current operating maturity, sustaining what has been demonstrated and working to continuously improve line performance over time.

Note: Tools and techniques have been described at various points throughout this standard. However, they should not be considered uniquely to the maturity level at which they were introduced. They can be used at any point in the manufacturing process where their application adds value.

Six Phases of Process Maturity

1) Process Standardization

Objective: Establish structure, discipline, and accountability in the workplace. Reduce process variation, highlight process waste, and create a baseline for training operators and implementing deviation management. Good standards will make deviations easy to identify.

Standardization reduces process variation by creating a common and repeatable way for employees to perform specific work activities. It helps to deliver a more consistent and predictable product output in alignment with customer expectations. Every facility must establish process standards where critical activities are performed to deliver products and services that meet customer requirements.

Applicable Methods/Tools/Techniques

Standard Work Instructions

Standard work instructions (SWIs) describe work that is highly specified as to content, sequence, timing, and outcomes. SWIs should address the work assembly sequence, handling of non-conformities, material supply requirements, and management of tools and documentation. At a minimum, SWIs must be prepared for manufacturing assembly operations.

Standards should include the warehouses (incoming and shipping), milk run, water spider, maintenance, and total productive maintenance (TPM) process once the assembly operations are completed.

Guidelines for preparing the SWI:

- Include safety and ergonomic elements.
- Define operational steps (elements) in detail including right-hand and left-hand movements.

- Include (a lot of) detailed pictures and other visuals.
- Descriptions should be clear and concise.
- Include customer-specific requirements and symbols.
- Photos should be large enough to clearly visualize the points being demonstrated or reinforced.
- Font size chosen for the SWI should allow everyone with 20/20 vision to comfortably read the document.

Plants can use their own templates if the content contains the elements in the following example:

Place an example of SWI here!

Standard Operation Sheet

The standard operation sheet (SOS) is used to document the number of operators and workflow among other relevant operating information. An SOS is required for every production line. A separate SOS must be available at the line information board (LIB) for every operator configuration. At a minimum, the SOS should display the following information:

- Planned cycle time.
- Production line layout.
- Work sequence.
- Position and motions of operators (with and without parts).
- Line feeding systems.
- Position of material stock/standard work-in-process (WIP).

Guidelines for Preparing the SOS:

- Align SOS numbering with SWI steps.
- Operator movement with parts, use solid arrows.
- Operator movement without parts, use dotted line arrows.

Place an example of SOS here!

5S Method

5S is a systematic approach for organizing, standardizing, and maintaining a clean and orderly workplace. The intent of 5S is to ensure all necessary materials, equipment, and information are available for efficient work execution. Every facility must have a documented 5S system at the production line and in all support functions that conform to the 5S standard. 5S standards must be defined for each area, integrated into area SWRs, become a working standard, and audited weekly as part of the LPA process. Deviations from the standard should be followed up immediately. Reoccurring issues must be addressed using the deviation management process.

2) Process Stability

Objective: Identify the causes of process abnormalities (e.g. deviations) contributing to process instability and use problem-solving to identify and permanently eliminate root causes so that process output is consistent and predictable over time. Stability is the foundation for sustainable improvements.

Process stability is achieved when process output (e.g. units per hour) is consistent and predictable over time. Stability can be assessed by using statistical process control (SPC), a run chart, or by monitoring hourly output on the LIB. The criteria for stability may look something like the following: *Number of units produced each hour within ±10% of the output average for 10+ operating shifts.*

Applicable Methods/Tools/Techniques

Production Line Output Chart

Each production line will have an output chart on their LIB to monitor the line's hourly output performance. This chart will be the primary source for triggering deviation management activities when significant variation from expected output performance occurs. Triggers for deviation management can include:

- Output performance beyond ±10% of the hourly targeted output.
- Output greater than ±3 Sigma upper and lower limits on the SPC chart.
- Run Chart output indicates patterns, shifts, or trends in the data.
- Overall equipment effectiveness (OEE) below its daily target.

Place an example of the product line output chart here!

Statistical Process Control Charts

Statistical process control (SPC) is a tool to assess, establish, and monitor process stability. SPC charts are trend charts with data-based control limits used to determine whether a process is in statistical control (e.g. stable). They can aid in maintaining process control (e.g. stability over time) by highlighting uncontrolled process variation for elimination. SPC is one of several options for assessing, achieving, and maintaining process stability.

Place an example of an SPC chart here!

Run Charts

A Run Chart is a tool to identify trends, shifts, or patterns in process output data that reveals abnormal or unstable process behavior over time. It's used to monitor process variation and highlight when unacceptable variation is negatively impacting process stability. Each plant must consider how they will determine process stability and their method for assessing, achieving, and maintaining a stable output over time.

Place an example of a run chart here!

Key Process Indicators

Key process indicators (KPIs) must be monitored daily by the production line teams, reviewed by management during Gemba walks, and discussed during performance reviews whenever unfavorable performance is evident. Action should be considered for any KPI below its expected target. KPIs to be monitored for daily output performance and long-term trends include:

- Production output.
- Overall equipment effectiveness (OEE).
- Scrap.
- First pass yield (FPY).
- Internal defect rate (IDR).

Jidoka

Jidoka is a method for enabling quick identification of deviations for immediate action to eliminate root cause. It may include stopping the process to prevent further defects from occurring or prevent additional losses. Quality is responsible for verifying the operational integrity and effectiveness of the Jidoka system. Jidoka limits must be reviewed quarterly with the objective of continuously lowering limits to highlight the next problems for eradication. If limits are not lowered during the quarterly review, justification for not doing so must be documented.

The objective is to continuously lower Jidoka limits to "one" defect so that action is taken immediately after one defect occurs. Clear work instructions shall be provided on how an operator must respond to a manual or machine Jidoka stop.

Place an example of a Jidoka chart here!

3) *Process Capability*

Objective: To ensure processes are meeting performance targets.

In the capability phase of process maturity, process stability has been demonstrated and output should be at 100% of expected targets. If not, the team must identify the sources of variation causing deviations from target and act to systematically reduce/eliminate those sources of variation until targeted performance is demonstrated. A process is capable when output performance consistently meets its targeted OEE. All OEE targets must be set at 85% with the expectation of achieving and sustaining 95% OEE. The OEE target must be reviewed quarterly with the objective of continuously increasing the target to improve operational efficiency. If limits are not lowered during the quarterly review, justification for not doing so must be documented.

Applicable Methods/Tools/Techniques

KPI Performance Charts with Targets

Process capability reflects the ability of a process to consistently meet performance targets. This means every KPI performance chart must have a target so that actual results can be compared with expected results. Any deviation from target should be visually evident on the chart and a topic for deviation management.

Place an example of a KPI chart here!

Performance Trends

Production line KPI trends should be reviewed daily by the product line team, during Gemba walks and periodic performance reviews. Appropriate actions should be taken (deviation management) whenever stagnant or negative trends are encountered.

Place an example of a trend chart here!

4) Process Control

Objective: To ensure a system is in place to maintain stable and capable processes. This will provide a baseline for driving continuous improvement initiatives.

Process control is the process of maintaining a stable and capable product output over time. This is accomplished through standard work routines (SWRs), layered process audits (LPAs), and deviation management activities. Every facility must have a set of SWRs and LPAs defining the activities and frequency for maintaining process control through all levels of management.

Methods/Tools/Techniques

Standard Work Routines

Standard work routines (SWRs) help to maintain process control and improve operational efficiency as part of the production operating system. Each level of management (e.g. Line Leaders, Supervisors, Production Managers, FF Managers, and Plant Manager) is expected to have SWRs to be performed at defined intervals.

Plants will define their own SWRs with the guidance (and agreement) of the Operations Manager. A guideline for preparing SWRs will be available to the plants for reference. This guideline must be considered when customizing plant-specific work routines. Any work routine in the guide that is applicable to the plant but not part of the plant specific work routines must be justified in writing.

SWRs should also be defined for all direct and indirect shop floor support functions such as quality, industrial engineering, maintenance, warehousing, and logistics.

Layered Process Audits

Layered process audits (LPAs) are performed to verify the production operating system and sub-systems are executed, functioning, and effective. Customer-specific requirements must be integrated into the LPAs to improve operational efficiency while satisfying customer expectations. A schedule of LPAs should be prepared to ensure they are being performed on a regular basis. Significant findings from these audits must be addressed in a timely manner with follow-up actions. LPAs typically include, but are not limited to, a review of the following systems, methods, or processes:

- Customer-specific requirements.
- Standard Work verification.
- 5S verification.
- Jidoka implementation/effectiveness.
- Visual management.
- Handling of scrap.
- Health, safety, and environmental.
- Technical cleanliness.
- Preventive maintenance/TPM.

Visual Controls

Visual control involves the communication of current performance data and trends in visual, easy to understand ways. This visual information is used to identify problems and triggering actions to maintain process control and drive improvements across the entire organization. Visualization supports standards and makes them visible to employees, customers, and visitors.

Visual Controls – Line Information Board

The line information board (LIB) is a visual control and a key element for deviation management. The production line leader is expected to take ownership of the board. Front-end and back-end production lines should have an LIB. At a minimum, the LIB is expected to contain the following information:

- Line name.
- Line responsible – name.
- Line status – using flags, lights, or other equivalent visuals with standards defined.
- Production line output chart/deviation chart (updated hourly).
- 4M/escalation chart (man, machine, method, material) – optional.
- OEE (updated every shift).
- Current problem-solving activities/manual 5-why.
- Action list.
- Escalation positions and phone numbers (if appropriate).

- Most recent return (one pager).
- Maintenance target/status.
- SOS chart.

Place an example of a Line Information Board chart here!

4M/Escalation Chart (Man, Machine, Method, Material)

The 4M chart is used to highlight and escalate line issues quickly to management for awareness and support. The chart must be filled in whenever a risk is identified or updated when a risk changes status during the shift. In parallel, an action can also be initiated to address the risk. The chart can be integrated into the production output chart.

Place an example of a 4M / Escalation chart here!

Deviation Management

Deviation management helps to maintain a stable and capable process through rapid identification and elimination of process deviations/abnormalities. Every facility must have a system for managing deviations that includes issue identification, analysis, action, and verification. The LIB is a key resource for deviation management. Deviation management should focus on the highest priority issues identified. Each plant must have a standard method for prioritizing actions to ensure the highest priority items are continuously being addressed.

Place an example of Deviation Management documentation chart here!

Deviation management can be triggered when:

- Output is not within expected performance limits.
- Manual or automatic Jidoka stops occur.
- KPI trends are not favorable.
- Problem reoccurrence has been identified.

Problem-Solving

The plant will use a systematic and structured approach to problem-solving such as A3, 8D, 5-Why, Kepner-Tregoe, Plan-Do-Check-Act, and Kata. The method selected should be suitable for the issue being addressed. All problem-solvers must demonstrate that the root cause was found, and the <u>permanent</u> corrective action (PCA) implemented was effective.

Each production line and support function must be actively and continuously engaged in problem-solving. Use of a scientific thinking approach to problem-solving is encouraged and required when applying Kata coaching. Evidence of current problem-solving activities must be at the LIB.

Paynter Charts
It's highly recommended that Paynter charts be used to record problem events and track the effectiveness of implemented countermeasures by monitoring for problem reoccurrences.

Daily Production Meetings

A communication structure must be in place to rapidly escalate daily production issues from the shop floor to the Plant Manager in a timely manner (e.g. within 2 to 3 hours). This structure should include:

- Meeting purpose/objectives.
- Location.
- Frequency.
- Time-period.
- A standard agenda.
- Expected attendees.

Communication meetings should occur daily (e.g. at the same time each day) at all levels of operations and focus on issues and risks requiring escalation to upper management. Consider quick, 15-minute stand-up meetings at the LIB or other production floor location. The expectation is everyone is prepared for the meeting with data and facts to keep it short, concise, and relevant.

Gemba Walk

Gemba walks increase management awareness and provide an opportunity to identify problems, discuss issues, offer support, and coach/develop people. Gemba walks also provide the Plant Manager and senior leadership an opportunity to strengthen relationships at all levels of the organization. The Plant Manager with members of staff (engineering, maintenance, quality, etc.) must tour 100% of the lines each week.

Visualize the process; go, see, and understand the issues and problems where value is created. Prepare a standard agenda for Gemba walks so management can make the most of their time on the production floor. Gemba walks should occur frequently (e.g. daily) and be included as part of a manager's SWRs.

Gemba Walk Tips

- Plan to perform daily Gemba walks on all shifts.
- Focus on deviation management activities.
- Review and engage teams doing problem-solving.
- Look for opportunities to coach and develop people.

- Recognize people for their good work to reinforce the right behaviors.
- Review all lines (stable and critical) at least once weekly.
- Senior leadership should reinforce the "value" of each employee's contribution to plant performance, at all levels of the organization.

Total Productive Maintenance

Total productive maintenance (TPM) is key to maintaining equipment reliability. It's intended to sustain overall equipment effectiveness (OEE) by preventing equipment deterioration while facilitating an environment of ownership between operators and their machines.

Each Plant must develop a strong, comprehensive TPM program focused on development of the line operator and technicians. This program at level 1 must teach team members the operating principles of their equipment and develop their ability to fully maintain the equipment at factory specifications. At level 2, the operator must be capable of addressing minor repairs and mid-level concerns, accepting some technician responsibilities.

TPM must be integrated into the SWR of those responsible for its execution. A good TPM program leverages lessons learned, best practice, and standard work procedures to improve maintenance efficiency and effectiveness. Lessons learned should be considered in future equipment design to create more robust and reliable equipment.

Every plant will identify their *critical spare parts* and work to maintain 100% of critical spare parts in inventory. The critical spare parts list should be updated whenever lack of a part in inventory causes unexpected and significant equipment downtime. Plants shall prepare a procedure for managing their spare parts inventory and audit to it periodically.

Poka-Yoke (Mistake/Error) Proofing

Error proofing is a technique to prevent mistakes and defects from occurring. It's applicable for all processes and should be kept simple. Look for opportunities to integrate error proofing into the problem-solving process to prevent problem reoccurrence. Use simple objects like fixtures, sensors, and warning devices to prevent human error. Poka-yoke checks for functionality and effectiveness should be integrated into SWRs or LPAs.

Performance Reviews

Performance reviews allow production teams to assess the effectiveness of existing process controls and continuous improvement activities. Weekly/monthly performance reviews must be held by management to review KPIs and other key metrics. These reviews must consider critical issues and performance trends to identify reactive and proactive countermeasures. Relevant data must be readily available, visualized, and presented to quickly communicate the information needed for problem solving and decision-making.

Topics to consider for the reviews include:

- Present and trending performance.
- Escalated and critical issues (safety, quality, productivity, costs).
- Current risks (4M).
- Status of strategic projects.
- People development issues.

5) Process Continuous Improvement

Objective: Employees autonomously identify and implement sustainable improvements and update process documentation (e.g. SWI and SOS) to establish a new standard (baseline) for performance and control. Areas of improvement include safety, quality, productivity, and cost.

Once process stability and capability have been verified and SWRs are in place to maintain process control, attention should turn to driving sustainable efficiency improvements. This can be facilitated in many ways, including observation, experimentation, project management, and change management.

Methods/Tools/Techniques – Waste Elimination

Process Mapping/Value Analysis

Process mapping is a technique for visualizing the sequence of activities within a process. The act of creating an "as-is" process map helps calibrate team members on the process's current state, and a value analysis of the results is likely to reveal opportunities to eliminate waste, reduce complexity, and decrease cycle time.

Functional support and production teams should map at least one process every 6 months and analyze the map for improvement opportunities. These opportunities should be prioritized, and actions taken to drive sustainable improvements using methods such as quality circles, Kaizen events, PDCA cycles, A3 thinking, and Kata projects.

Kaizen Events

Kaizen events are short duration improvement activities with a specific, targeted objective. These events typically result from Gemba walks, escalated issues, performance reviews, process mapping, value stream design (VSD), and other similar activities. Kaizen events should be planned and performed as needed to drive process improvements. They are conducted by area team members and support personnel. In a mature system, Kaizen becomes part of the culture and occurs continuously; it's no longer an event.

Kata Projects/PDCA

Kata is a PDCA-based method that uses an experienced coach to develop an employee's problem-solving skills. Kata takes a systemic and scientific thinking approach to eliminate obstacles to achieving a target condition for improved performance. Kata is best employed on non-critical but important improvement opportunities since time is required to develop the Improver. It should be used at the discretion of the management team as a problem-solving and employee development technique.

Change Management

Change management is a critical process for validating and institutionalizing process improvements. It involves updating working documents, operational standards and obtaining employee commitment to integrate process changes into their daily work routines. Every organization must have a documented change management process that is audited periodically (e.g. LPA) for efficient and effective execution. Sustainable improvements result from an effective change management process.

Self-Directed Work Teams

A self-directed work team (SDWT) is a group of people, usually company employees, who use their cross-functional skills, experience, and talents to work toward a common purpose or goal without traditional management oversight. These teams are typically formed to drive specific company initiatives and strategic projects that should be reviewed periodically for progress to planned performance.

Jishuken Workshops

Jishuken is a self-study activity to identify, analyze, and drive improvement actions in a specific area through Kaizen events. Jishuken workshops should be performed on an as-needed basis. It's highly recommended that workshops are performed at least once per quarter to drive performance improvements where needed. It's also important to conduct periodic workshops to maintain and develop in-house capability to execute workshops effectively.

Employee Idea Management

Employee idea management (EIM) is a way for facilities to encourage and capture employee ideas for consideration, implementation, and recognition.

6) Process (Value Stream) Optimization

Objective: To enhance the flow of information and material throughout the value stream by eliminating obstacles to flow.

Value stream optimization requires mapping current information and material flow through a product family's value stream to highlight obstacles to a steady, uninterrupted flow. Obstacles to flow increase process lead time. Improving flow can reduce lead time and facilitate the reliable delivery of products and services to customers. Opportunities for improving flow can be achieved by:

■ Reducing scrap and rework.
■ Improving equipment reliability.

- Reducing changeover time.
- Balancing operator workload.
- Reducing operator movements.
- Reducing work-in-process.

Eliminating obstacles to flow starts with identifying the gaps between the current and ideal (future) state within a product family's value stream. An improvement plan can then be developed to methodically remove the gaps, incrementally improving process performance.

Methods/Tools/Techniques

Pull and Flow

Manufacturing efficiency is driven by the steady flow of material through the production process. A pull system is a lean technique used to regulate production inventory by only producing product upon customer demand. The ideal state is to produce and deliver a specific quantity of good parts, upon request. This helps to reduce overhead and minimize inventory costs. Every facility must strive to improve flow and operational efficiency by eliminating flow disruptors and reduce WIP, buffers, and finished goods inventory. The following tools and techniques can assist with this objective.

Value Stream Mapping/Value Stream Design

Value stream mapping (VSM) and design helps teams visualize improvement opportunities between the current and ideal state of process flow. Every facility should prepare a flow improvement plan for one of their product family value streams every 6 months and actively drive improvements to the plan, periodically updating the current state map based on significant improvements to flow. It's important that each facility maintains and develops internal expertise to drive flow improvements through the active development, deployment, and management of value stream improvement projects.

Vendor Managed Inventory

Vendor managed inventory is an inventory management and order fulfillment model where the product buyer provides information to the vendor who takes full responsibility for maintaining an agreed to material inventory, usually at the buyer's consumption location. This business model helps avoid out-of-stock issues and reduces supply chain inventory. It can be used to regulate or reduce inventory and corresponding costs. Plants should consider the potential benefits of pursuing VMI for their location.

Kanban

Kanban is a visual scheduling system used to improve manufacturing flow and efficiency by controlling inventory levels in production. It should be used, as needed, to maintain and control flow. In non-manufacturing disciplines, it can be used to move (pull) tasks through a department (using a visual board).

Heijunka

Heijunka, a Japanese word for "leveling," helps organizations regulate workload demand by balancing the type and quantity of production output over a fixed time. This improves operational efficiency by creating a more stable and predictable process workflow. The Heijunka box is a tool to level the pace of production assembly when customer demand fluctuates significantly, causing disruptions in process flow.

Supermarkets/First In First Out

A pull system limits the amount of inventory in process. A supermarket is a tool to control the amount of work-in-process and helps to maintain process flow within specified parameters. Supermarkets can work in concert with Kanban to support a pull system. As parts are withdrawn from the supermarket, a part count dropping below a minimum threshold triggers the Kanban process to request more parts upstream of the supermarket. Parts should be withdrawn from the supermarket in a first in first out (FIFO) sequence to maintain system integrity for problem-solving and defect containment. Consider using supermarkets/FIFO to maintain process flow within the value stream.

Strategic Projects

Every facility must identify and articulate their fundamental activities for process control and continuous improvements. This includes identifying one or more strategic projects for each focus factory annually. Examples of strategic projects can include changes in the plant layout to improve material flow, a plant-wide focus on material handling, and optimizing a product family's value stream flow. Each strategic project will be reviewed at least monthly to evaluate progress, assess effectiveness, and make appropriate adjustments based on data and information collected.

Appendix 3 – Standard Work Routines

Example of Standard Work Routines

STANDARD WORK ROUTINES					Suggested Starting Frequency							
Activity	Reference Documents	Line	Plant	Description of Routines	Shift	Daily	2X/W	W	2X/M	M	Q	A
5S Method		X		**5S Compliance** – perform a quick 5S compliance check of the production line	LL	S			FM	PM		
			X	**5S Compliance** – perform a 5S assessment (shop floor, warehouse, support operations, administration, etc.)				S	FM	PM		
				Performance Review – review 5S functionality and effectiveness; review 5S scores for stability and trends						All		
				5S Roadmap – review execution of plant 5S roadmap to plan. Address any deviations						All		
Continuous Improvement			X	**Continuous Improvement Activities** – perform a random review of plant continuous improvement activities (e.g. Kaizen events, Jishuken workshops, etc.)				S				
		X		**Process Maturity Assessment** – Confirm a line maturity assessment was performed and improvement actions are in process. Evaluate effectiveness of completed and current activities. Coach as required					S	FM	PM	
				Performance Review – Review status of plant improvement projects						All		

Category	Standard Work Routine	X	LL	S	FM	PM	All
Environmental, Health & Safety (EHS)	Confirm safety policy and procedures are being followed	X					
Electrostatic Discharge	**Electrostatic Discharge Line Compliance** – Confirm **Electrostatic Discharge** line compliance	X	LL	S			
	Electrostatic Discharge Compliance Review – Review **Electrostatic Discharge** audit results (plant level)						All
	Line Deviation Management – Deviations from standards and procedures (e.g. Standard Work, Leader Standard Work, Key Process Indicators, etc.) are being identified, properly addressed, and closed in a timely manner	X	LL				
Deviation Management	**Line Confirmation** – Confirm Deviation Management is performed, functional, and effective. (e.g. Standard Work verification, 5S, Jidoka, Key Process Indicators)	X		S	FM	PM	
	Deviation Management Plant Confirmation – Confirm deviations are found and addressed by plant staff (Gemba walks, Layered Process Audits, performance reviews, etc.)				FM	PM	

Example of Standard Work Routines

STANDARD WORK ROUTINES					Suggested Starting Frequency							
Activity	Reference Documents	Activities		Description of Routines	Shift	Daily	2X/W	W	2X/M	M	Q	A
		Line	Plant									
Gemba Walk		X		Perform Gemba walk of production line		S						
			X	Perform general or area-specific Plant Gemba walks								
				Confirm Gemba walks are efficient, effective, and actions identified are prioritized and followed up						All		
Jidoka		X		Assess Jidoka functionality and effectiveness at the line			LL	S				
				Assess Jidoka functionality and effectiveness randomly throughout the plant				Q	FM			
		X		Confirm production line maintenance activities are performed to schedule. Escalate if necessary.	SSL		S		FM	PM		
Preventive Maintenance			X	Confirm facilities maintenance activities are performed to schedule. Escalate if necessary.			S	S	FM	PM		
				Review compliance to maintenance schedules; address reasons for deviation. Assign corrective actions.								

Category	Task			LL	S	FM	PM
Problem Solving	Confirm line problem-solving activities are occurring, effective, and timely.	X			S	FM	
	Confirm plant problem-solving activities are occurring, effective, and timely.		X		S	FM	PM
	8D/A3 – Review quality of customer and internal problem-solving activities	X		LL	Q	FM	PM
Process Control	Confirm line output is consistent (stable) and within target (capable). Confirm instability and deviations from target are being prioritized and addressed	X		LL			
	Randomly confirm line output is consistent (stable) and within target (capable). Confirm instability and deviations from target are being prioritized and addressed				S	FM	PM
	Confirm Key Process Indicators are stable or improving. Confirm issues needing escalation to management are occurring	X		LL			
	Randomly confirm Key Process Indicators are stable or improving. Confirm issues needing escalation to management are occurring				S	FM	PM
Shift Start-Up and Changeover	Conduct a shift changeover and production start-up/re-start review at appropriate times	X		LL			
	Confirm that shift changeover and production start-up/re-start reviews/meetings are being conducted	X			S		FM

Example of Standard Work Routines

STANDARD WORK ROUTINES	Reference Documents	Activities Line	Activities Plant	Description of Routines	Suggested Starting Frequency Shift	Daily	2X/W	W	2X/M	M	Q	A
		X		**Material Handling** – Confirm material is properly handled within the production line	LL	SL		FM	PM			
Materials Management			X	**Material Handling** – Confirm material is properly handled within the plant (e.g. moved, stored, sorted, reworked, and refabricated)				FM		PM		
				Material Handling – Confirm material management system is working and effective (non-conforming, suspect, rework, reprocessing, and sorting)				S	FM	PM		
				Review open and overdue issues list		PM						
Poka-Yoke		X		Confirm line Poka-Yokes are working and effective				LL		S		
			X	Confirm plant Poka-Yokes are working and effective						FM		
Standard Work Confirmation		X		Standard Work Instruction confirmation (minimum: all operators 1X per week)	LL							
		X		Standard Work Instruction confirmation (minimum: all lines 1X per week)		S						
		X		Confirm Standard Work is being performed by next level down			S	FM				
				Standard Work confirmation – Random check					FM	PM		

Category	Activity								
Strategy Review	Progress review of strategic projects	X						FM	Staff
	Review 5S roadmap progress	X						FM	Staff
Technical Cleanliness	Perform line technical cleanliness check		LL						
	Confirm technical cleanliness checks are being performed by next level down				S				
	Perform line cleanliness check	X						FM	PM
	Perform plant cleanliness check							FM	PM
	Particle monitoring						Q		
	Perform annual Technical Cleanliness assessment								Q/PM
Visual Controls	**Line Information Board (LIB)** – Confirm LIB compliance to standard	X				LL	S	FM	
	Plant – Perform Visual Controls Audit					S	S	FM	PM

PM = Plant Manager

FM = Factory Manager

Q = Quality

S = Supervisor

LL = Line Leader

Appendix 4 – Tips for Shop Floor Monitoring

- Make a point to visit the shop floor daily, if only briefly, to greet and talk to employees at the front lines of operations.
- Talk to people directly. Express interest in what they are doing. Ask meaningful and probing questions.
- While on the shop floor, take time to mentor and coach people, when the need is apparent.
- Review output performance. Check for patterns and trends in key process indicator (KPI) data.
- Use visual aids to enhance shop floor monitoring efficiency.
- Look for unusual behaviors, process abnormalities, slow processes, equipment micro-stops, and opportunities for improvement. Follow up with needed actions.
- Perform a layered process audit.
- Make time on the shop floor part of an employee's standard work routine.
- Spend time (do a deep dive) at one process to enhance your knowledge, know-how, and know-why.
- Review on-going problem-solving activities/coach employees.
- Use mobile devices to access data from anywhere.

Appendix 5 – Single-Minute Exchange of Die (SMED) Workshop

Single Minute Exchange of Die

Single-Minute Exchange of Die Process

Changeover is the process of converting a production line or machine from running one product or variant to another. The following steps can be used to conduct a single-minute exchange of die (SMED) workshop to reduce product line changeover time when switching from one product variant to another. Changeover time is the time required to change a piece of equipment from producing the last good part of one model to the next good part of a subsequent model in a production line. If there are multiple equipment changeovers on a production line, the changeover time of the production line will be that of the longest equipment changeover time.

The objective of this workshop is to reduce the changeover time to the minimal time possible, while maximizing process run time. The simplified process will then be standardized for future application.

Step 0 – SMED Workshop Organization

- Select an improvement opportunity and team participants.
- Observe a changeover. Video record the process.
- Take notes on any improvement activities (without restrictions).
- Note changeover time to baseline the process.

Step 1 – Analyze the Selected Changeover Process

- Determine workflow and process steps.
- Note all movements (Spaghetti Diagram).
- Visualize the process (Gantt Chart).
- Identify waste opportunities for elimination (seven process wastes, 4M+E).

Step 2 – Separate Process Steps into Internal and External Activities

- Identify and document ***internal*** activities (activities that must be performed when machine is stopped).
- Identify and document ***external*** activities (activities that can be performed while the machine is running…no downtime!).

Step 3 – Convert Internal Activities to External Activities

- What internal activities can occur while the equipment is running (e.g. tools are staged and ready to use)?
- What activities can only be done when the machine is stopped (e.g. die installation)?
- What activities can be completed when the machine is back in production (e.g. tool cleaning, die storage)?

Step 4 – Optimize Internal Changeover Activities

- Apply the "ECRS" method: Eliminate, Combine, Rearrange, and Simplify to optimize all activities.
  - ➢ Can an activity be eliminated?
  - ➢ Can an activity be combined with another activity?
  - ➢ Can setup time be reduced by rearranging the sequence?
  - ➢ Can the setup time be simplified?

Step 5 – Optimize External Changeover Activities

- Apply the "ECRS" method: Eliminate, Combine, Rearrange, and Simplify to optimize all activities.
  - ➢ Can an activity be eliminated?
  - ➢ Can an activity be combined with another activity?
  - ➢ Can setup time be reduced by rearranging the sequence?
  - ➢ Can the setup time be simplified?

Step 6 – Standardize the Process

- Document standards to prepare and conduct changeover activities.

Note: The standard becomes the basis for continuous improvement.

It's important to baseline the current changeover process before starting. If possible, observe and record the time of several changeover events to get a full picture of process variation and improvement opportunities. Consider observing the same changeover event on different shifts. If a standard changeover process already exists, but needs to be improved, ensure the current standard is being followed before measuring cycle times.

Appendix 6 – Jishuken Workshop Methodology

Overview

Workshops don't just happen. Detailed planning, proceeded by disciplined execution, leads to successful workshop outcomes. The following structure will serve as a framework for conducting efficient and effective workshops. Review the list to identify relevant actions in preparation for a specific type of improvement event. It can be customized for an organization's unique practices, processes, and procedures while continuously being enhanced over time. Let's step through the life cycle phases of a typical improvement workshop.

Workshop Pre-Work (2 to 4 Weeks Before)

Description: Pre-work activities set in motion the actions necessary for day 1 of the workshop. These activities include production line selection, establishing workshop objectives, preparing an agenda, and gathering historical data and information to support workshop execution. The objective is to review the following checklist items and determine which apply for the type of workshop being organized.

Pre-Workshop Activities Checklist

- Select targeted product line(s) for improvement.
- Define workshop objectives/goals/targets.
- Notify manufacturing teams of their needed support during the upcoming workshop (e.g. quality, maintenance, engineering, supply chain, etc.).
- Manage workshop logistics:
  - Location.
  - Day and times.
  - Room (size, reservation, and arrangement/layout).
  - Projector/screen.
  - Flip charts/whiteboard/markers.
  - Food – meals/snacks/drinks.
  - Materials (paper, tape, rulers, markers, scissors, etc.).
  - Handouts.
  - Attendance list.

- Prepare a workshop agenda:
  - Activities and timing.
  - Number of days.
  - Lunch, breaks.
  - Evening event.
- Identify a local workshop leader and coach.
- Invite appropriate participants (e.g. management, support functions, maintenance, line leaders, and supervisors, etc.).
- Request confirmation of participation.
- Confirm 5S is implemented on the lines targeted for improvement. If not, contact your coach.
- Verify standard work instructions (SWIs) are available and followed on the lines targeted for improvement. If not, contact the responsible line leader.
- Schedule the workshop, distribute email, and communicate meeting logistics.

Workshop Preparation Activities Checklist (1 to 3 days Before)

- Document planned cycle time.
- Determine available working time per shift (e.g. 480 minutes per 8 hour shift minus lunch, breaks, meetings, training, etc.).
- Record number of changeovers per shift (typical).
- Determine time to perform changeovers (average).
- Record the number of operators that can work on the line.
- Note the number of product variants manufactured on the line and their differences.
- Check production schedule to confirm line will be running during workshop.
- Collect historical data:
  - Key performance indicator (KPI) trend data/graphs (prior 3 weeks of history).
  - Production line output performance.
  - Overall equipment effectiveness (OEE) performance (per machine and overall).
  - Line quality performance (First Pass Yield, scrap, rework).
  - Equipment downtime data (planned and unplanned).
- Determine process bottlenecks.
- Understand line layout and operator workflow for all operator configurations.
- Confirm standards are implemented and compliant (5S, SWI, standard operation sheet [SOS], control plan, parameter cards, etc.).
- Confirm product line stability (consistent output/statistical process control and capability (output consistently in spec)).
- Confirm process control is being exercised (e.g. standard work routines [SWRs], layered process audits (LPAs), Gemba walks, performance reviews, SWI verification, deviation management, etc.).
- Confirm:
  - Food and drink orders (lunch and breaks).
  - Room availability.
  - Workshop equipment availability (projector, flip charts, white board, markers, electrostatic discharge [ESD] jackets and straps, hair nets, etc.).

- Number of people attending.
- Planned workshop events.
■ Communicate:
  - Agenda topics.
  - Workshop site location.
  - Room location.
  - Meeting times.
  - Recommended hotels (and map to hotels).
  - Directions to/from airport.
  - Attendees list.
  - Dinner plans.
  - Transportation.

Workshop Activities (Day 1)

The workshop overview occurs at the beginning of the workshop and typically covers the following topics:

- **Workshop agenda** – review the agenda with management and participants. When presenting the agenda, cover the approximate timing of workshop activities, frequency, and length of scheduled breaks and lunch. This task may also include a review of guidelines and ground rules for the workshop. Make minor real-time changes to the agenda (e.g. typically timing), if necessary.
- **Workshop scope/objectives/targets** – clearly communicate the quantitative target(s) and expected outcomes of the workshop. These items serve as the basis for assessing workshop success. A project charter can be used to document workshop objectives and targets, among other details such as workshop scope, participants, and known constraints.
- **Participant introduction** – introduce the core team members of the workshop, in addition to the extended support team members from the different functions and departments, who are expected to be available upon request.
- **Training modules** – during the workshop, there may be opportunities to introduce or enhance awareness of topics relevant to the activities being discussed. As a result, training material can be inserted into the agenda at strategic points during the workshop to establish and confirm understanding of the method, tool, or technique before exercising it. This is done to present information at the point where it's most useful and can provide maximum impact on workshop activities and outcomes. The workshop experience can be enhanced when the right information is presented at the right time.
- **Go, see, observe to understand** – each workshop will start with time at the Gemba to observe the process. This activity is intended to heighten awareness and increase familiarity of workshop participants with the production line, operator work routines, equipment stability, and man–machine interactions. It's an opportunity to witness the process in action, noting the work rhythm of the line, any abnormalities and potential opportunities for improvement (OFI).

■ **Define the current state** – once the overview and observational activities are completed, it's time to establish the process baseline and document the operation's current state. This requires collecting data and information from actual process performance. Upon collection, the data and information should be visualized to create awareness and used as a baseline to measure process improvements. If a project charter was prepared, it should be updated to reflect the current state.

Workshop Activities (Day 2+)

■ **Process improvement** – during the workshop, various actions and OFI should have been documented for further consideration and implementation. It's important to implement as many of the immediate or short-term opportunities as possible so that you can evaluate their impact on the targeted objectives. Now's the time to review and prioritize these items, implementing quick win opportunities within the remaining timeframe of the workshop. An action plan can be prepared for the short-term (<1 week) and long-term (>1 week) actions as part of continuous improvement.

It may be appropriate during the improvement process to simulate or model proposed improvements to evaluate their effectiveness and quantify their impact. Data can be collected to verify the benefits of implemented solutions on process performance. A before versus after metrics chart can be prepared to show the value of improvements made during the workshop. Strive to implement simple and cheap solutions.

■ **Process control and sustainability** – Once improvements have been implemented and verified effectively, it's time to ensure the gains achieved are sustained. This is the purpose of process control: to maintain a stable and capable process over time relative to the latest operational standards. As the standards change, the activities to maintain process control need to reflect the elevated state of operational performance. Typical activities required to maintain the latest process standards may include updating:
 – process documentation and software.
 – standard work routines.
 – visual controls.
 – performance targets.
 – guidelines.
 – training materials.
 – layered process audits.

These and other activities should be reviewed after each change to ensure all process enhancements are properly maintained and controlled to prevent the loss of incremental improvements over time.

Workshop Closure Activities

It's important to communicate what was accomplished during the workshop to key stakeholders. This provides an opportunity to showcase the value of a well-executed event and promotes the individuals who contributed to its success. Effective workshop closure should include the following:

- **Management presentation** – at the end of the workshop, it's important to have management participation when reviewing workshop results. This not only reinforces the value of the event but allows team members to recognize the importance of these workshops in driving continuous improvements. The management presentation should be short (around 20 to 30 minutes). It typically consists of a series of four to five slides (instead of an agenda) which communicate the following:
  - Workshop objectives (e.g. project charter).
  - A Pareto chart of top issues or concerns found.
  - A list of short- and long-term actions identified (e.g. action plan).
  - Impact of actions implemented during the workshop (metrics chart – before/ after).
- **Improvement (action) plan** – before ending the workshop, it's important to assign a single individual to oversee the closure of all short- and long-term open issues. In addition, each action should be assigned an owner and due date so that appropriate follow-up can be achieved. Open actions can be closed through verification of effectiveness, terminated (if no longer feasible), or canceled in light of unexpected circumstances.
- **Follow-up meetings** – hold periodic reviews to ensure all relevant action items are properly closed.
- **Participant recognition** – take time to thank all participants for their time and efforts in the presence of management.
- **Best practices/lessons learned** – a summary of lessons learned and best practices identified during the workshop should be used to upgrade existing processes and procedures in the spirit of continuous improvement.
- **Participant feedback** – obtain participant feedback and suggestions for improving future workshops.

Post-Workshop Activities

Workshop activities typically extend well beyond the time allotted for the workshop. Responsible individuals should be identified and assigned to follow-up with the longer-term action items. Whenever possible, any changes to an existing process should be verified as effective before closing an action item. When an action has been completed and proven effective, consider sharing it with other areas and location that can benefit from it. Toyota calls this Yokoten.

Yokoten is the process of sharing best practices and lessons learned across an organization. The expectation is to consider vertical and horizontal deployment of lessons learned within an organization; vertical meaning across functional departments and operational support areas (e.g. sales and marketing, R&D, manufacturing, and service). Horizontal meaning peer-to-peer sharing of what you learned. Yokoten is not a copy-and-paste mentality, it's a copy-and-improve opportunity since all areas don't function the same way.

Appendix 7 – Acronyms and Terms

AGVs   Automated guided vehicles.
AI   Artificial intelligence.
ASRS   Automated storage and retrieval systems.
BOM   Bill of materials.
Cpk   Capability analysis.
CRB   Change review board.
DE&I   Diversity, equity, and inclusion.
DfM   Design for manufacturing.
DFSS   Design for six sigma.
DICOV   Define, identify, characterize, optimize, and validate.
DM   Deviation management.
DoE   Design of experiment.
DSFM   Daily shop floor management.
EHS   Environmental, health, and safety.
EPA   Electrostatic discharge protected areas.
ERP   Enterprise resource planning.
ESD   Electrostatic discharge.
FMEA   Failure mode and effects analysis.
FTA   Fault tree analysis.
GR&R   Gauge repeatability and reproducibility.
IoT   Internet of things.
JIT   Just-in-time.
LIB   Line information board.
LPA   Layered process audits.
MRO   Maintenance, repair, and operations (materials).
MRP   Materials requirements planning.
MTBF   Mean time between failures.
MTTR   Mean time to repair.
NIOSH   National Institute for Occupational Safety and Health.
NUD   New, unique, and difficult.
OEE   Overall equipment effectiveness.
OEM   Original equipment manufacturer.
OFI   Opportunities for improvement.
OHL   Occupational hearing loss surveillance.

OSHA	Occupational safety and health administration.
PDCA	Plan, do, check, act.
PDSA	Plan, do, study, act.
PEL	Permissible exposure limit.
PFEP	Plan for every part.
PPE	Personal protective equipment.
ROI	Return on investment.
SCM	Supply chain management.
SIPOC	Supplier, input, process, output, customer.
SMED	Single-minute exchange of die.
SOS	Standard operating sheet.
SWI	Standard work instruction.
SWR	Standard work routines.
TIP	Tactical implementation plan.
TPM	Total productive maintenance.
VMI	Vendor managed inventory.
VSD	Value stream design.
VSM	Value stream mapping.
WIP	Work-in-process.

Index

Pages in *italics* refer figures.

Printed in the United States
by Baker & Taylor Publisher Services